Poultry farming & feed formulations: Nutrition, methods & strategies of feeding

Poultry farming & feed formulations: Nutrition, methods & strategies of feeding

Editor

Almiro Salazar

scitus academics

Poultry farming & feed formulations: Nutrition, methods & strategies of feeding

Edited by **Almiro Salazar**

ISBN: 978-1-68117-624-6
Library of Congress Control Number: 2016947443

616, Corporate Way, Suite 2, 4766,
Valley Cottage, NY 10989

www.scitusacademics.com

Printed in United States of America on Acid Free Paper

Preface

Poultry farming means 'raising various types of domestic birds commercially for the purpose of meat, eggs and feather production'. The most common and widely raised poultry birds are chicken. Chickens, turkeys, ducks, and geese are of primary importance, while guinea fowl and squabs (young pigeons) are chiefly of local interest. Whatever the production system, all management procedures with adult stock – during incubation and hatching, brooding of young chicks, and rearing of young meat and layer stock – should focus on meeting the birds' physiological requirements at all stages of life by providing an ideal physical environment, minimizing exposure to disease, meeting the birds' behavioral and social needs, and providing them with clean water and good-quality feed that satisfies their nutrient requirements. Animals eat to acquire the energy and building materials that they need to live and grow. Animals use energy to perform normal body functions such as breathing, walking, eating, digesting, and maintaining body temperature. Nutrients provide poultry the energy and material needed for the development of bone, flesh, feathers, and eggs. Water is often overlooked, but it is one of the most important nutrients. An animal can live without food longer than it can live without water. Water plays an important role in the body of an animal. Water softens feed and carries it through the digestive tract. Carbohydrates (compounds with carbon, hydrogen and oxygen) are an energy source for animals and make up the largest portion of a poultry diet. Proteins are used in the construction of body tissues such as muscles, nerves, cartilage, skin, feathers, beak, and so on. Egg white is also high in protein. Minerals play a role in bone formation, but minerals are also needed for several other important functions, including formation of blood cells, blood clotting, enzyme activation, and energy metabolism and for proper muscle function. Vitamins are a group of organic compounds that poultry require in small quantities. Despite the low requirement levels, vitamins are essential for normal body functions, growth, and reproduction. If poultry are to achieve their genetic potential for meat or egg production, they need an environment that meets their physiological requirements. This includes: a suitable physical environment in terms of temperature, humidity, air movement and the surfaces on which they live; adequate food and water; minimal exposure to disease organisms; and avoidance of exposure to stress resulting from the physical and social environment.

Poultry Farming & Feed Formulations: Nutrition, Methods & Strategies of Feeding presents topics on how appropriate diets can be formulated and how feeding programs can be integrated into poultry production system.

Table of Contents

CHAPTER 1

Comparison of Indoor and Outdoor Bioaerosols in Poultry Farming

Elżbieta Lonc and Kinga Plewa

Department of Microbial Ecology and Environmental Protection, Institute of Genetics and Microbiology, Wrocław University, Poland

1. INTRODUCTION

Intensive poultry production, implying large densities of animals in small areas, is a significant source of air pollution which may constitute a considerable health hazard to the birds, farmers and those living in the proximity of the farm (Lonc & Plewa, 2009). On the other hand, the spread of bioaerosols on the outside of animal housing may result in local or even more extensive environmental pollution (Bakutis et al., 2004).

Under commercial production the airborne particles will contain a mixture of biological material from a range of sources. The chickens produce large amounts of dust as a result of epithelial desquamation, as well as from feed, manure, faeces and litter (Matković et al., 2009). This dust consists of a variety of airborne particles of biological origin, i.e. bacteria, fungi, endotoxins (lipopolysaccharide, LPS) of Gram-negative bacteria, 1.3-beta-glucan of fungi, fungal spores and mycelium fragments. Hence, a more descriptive term for these airborne particles is bioaerosol in which the microorganisms can occur either as liquid droplets or as dry particles [Dutkiewicz, 1987; Matković et al., 2009; Nevalainen, 2007]. In specific conditions, bioaerosols may show pathogenic, toxic or allergy-causing effects. The particles in a bioaerosol are generally 0.3 to 100 μm in diameter; however, the respirable size fraction of 1 to 10 μm is of primary concern. Bioaerosols, ranging in size from 1.0 to 5.0 μm, generally remain in the air, whereas larger particles are deposited on surfaces (Srikanth et al., 2008).

Bioaerosol may contain representatives of Gram-positive bacteria: *Corynebacterium, Staphyloccocus, Streptococcus, Micrococcus, Pantoea* and *Sarcina* (Siemiński, 2001). Their presence in large numbers may present a significant immunological challenge to the human respiratory system. In dust are suspended also endotoxins (lipopolysaccharide complex - LPS) associated with the outer membrane of Gram-negative pathogens, such as *Escherichia coli, Salmonella, Shigella, Pseudomonas, Neisseria* and*Haemophilus influenzae.* LPS is composed of two major parts, the hydrophobic lipid A and the hydrophilic polysaccharide part (commonly called the "O" region). Most biological effects of LPS are due to the lipid A part, however O-region plays an important role in effective colonisation of host tissues. Inhalation of organic dust contaminated by endotoxins may cause chronic bronchitis and inflammatory reaction in the lungs (Bakutis et al., 2004, Schierl et al., 2007, Pomorska et al., 2009).

As with bacteria, the fungi present in the poultry dust bioaerosols may be derived from soil, dust feed and litter, but to a lesser extent from the birds themselves. Fungi are ubiquitous in all atmospheres. In general, both outdoor and indoor atmospheres are dominated by representatives of the genera*Cladosporium*, *Penicillium*, *Aspergillus, Alternaria* and by yeasts and Mycelia sterilia. *Cladosporium* is nearly always the dominant fungus in outdoor as well as indoor atmospheres. The abundance of the other fungi varies with the season and place. In relation to outdoor environments, indoor atmospheres typically display lower diversity (Araujo & Cabral, 2010). Long-term or repeated exposure to high concentrations of airborne fungal spores is recognised as contributing to the decline in lung function and allergic diseases such as asthma and allergic alveolitis known as farmer's lung disease (Crook et al., 2008).

Literature data usually pertain to the air biopollutant concentration inside the poultry houses. Much less is known about the relationships between the indoor and outdoor biological pollution, as well as about the spreading of indoor bioaerosols in the surroundings of the farms.

The aim of the study was to assess the influence of microbiological air contamination in the intensive poultry breeding, both inside and outside farms. The comparative quantity and quality analysis concerns bacteria as well as fungi isolated from the air samples taken during two seasons.

2. MATERIALS AND METHODS

Seasonal sampling was conducted in the summer of 2009 and spring of 2010 in two (I and II) poultry houses on family farms located near Wrocław in Lower Silesia, Poland (Fig. 1.)

Both farms were accommodated to 18 000 and 23 000 broilers, with the density of 16-17 chicken per square meter. The broilers were kept on the rye straw deep litter in buildings equipped with mechanical ventilation (inlet and outlet ventilators), heating with a central thermogen and artificial lighting with regularly distributed bulbs.

Air samples were taken using a MAS-100 air sampler (Merck KgaA, Darmstadt, Germany) which is representative of the new generation impactor samplers and is frequently used for indoor and outdoor sampling. These instruments are based on the principles described by Andersen and aspirate air through a perforated plate, which results in impaction of particles from an airstream onto the surface of agar medium. The speed of air flow through the sampler was about 11 m/s, air volumes were 5-200 litres (depending on the expected contamination level) and the sampling rate was 100 l/min. Indoor and outdoor samples were collected in the poultry biozone during the fattening period. The biopollutants were determined on the basis of airborne bacteria and fungi. The samples for each group of bacteria and fungi were taken at the central point of poultry houses 1.3 m from the ground level. The emission level outside the farming objects was determined similarly, i.e. 1.3 m with sampling points situated 10 m, 50 m and 100 m away from the farming buildings. At the same time both humidity and temperature were measured with a termohigrometer (Label).

Microbiological studies of the air samples were used to determine the number of mesophilic bacteria, *Enterobacteriaceae* representatives, mannitol+ staphylococci, *Salmonella sp.* and mould fungi. The mesophilic bacteria were isolated with the use of TSA agar (BioMérieux).

Enterobacteriaceae families were determined with the use of VRB medium (BioMérieux). The estimation of *Salmonella sp.* was done on SS agar (BioMérieux). Mannitol salt agar (BioMérieux) plates were inoculated for culturing and counting *staphylococci.* Mould fungi were determined with the use of Sabouraud (Merck) medium. Colonies were counted after 48 h of incubation at 37^0C for bacteria and after 5 days at 26^0C for moulds and subsequently the colony-forming units (CFU) were determined. Quantitative results were expressed in CFU/m3, i.e. colony forming units in 1

m^3 of the examined air and the total microbial count was corrected using the conversion formula devised by Feller:

Figure 1. Poultry houses I and II; the sampling sites outside the poultry house at the distance of 10 m, 50 m and 100 m from the farming object respectively, as well as at the center of building.

$$Pr = N\,[1/N + 1/N\text{-}1 + 1/N\text{-}2 + \ldots + 1/N\text{-}r+1]$$

where:

N = 400 (number of holes in perforated lid of the sampler)

r - number of CFU counted on Petri dish

Pr - statistically corrected total count of bacteria/moulds in tested air volume

Bacterial species were identified on the basis of gram staining, microscopic morphology, oxidase and coagulase activity, catalase test results and metabolic properties according to standard procedures described in Bergey's Manual of Determinative Bacteriology (2001). The following commercial systems were used: API 20E (BioMérieux, France) for enteric gram-negative organisms; API 20 NE (BioMérieux) for fastidious and nonfermenting gram-negative organisms; API Staph (BioMérieux) for gram-positive staphylococci, API 20C AUX (BioMérieux) for identification of yeasts, Slidex - Strepto Kit (BioMérieux) for the identification of Lancefield A,B,C,D,F et G group antigens of streptococci and Slidex Staph Plus (BioMérieux) to detect clumping factor, protein A and group-specific antigen bound to the *S. aureus*. Moulds colonies were identified on the basis of colour, texture, topography of the culture surface, smell of the colony, colour of the reverse of the colony and the presence of the diffuse pigment. Microscopic features of the fungal colonies, i.e. the presence of macroconidia and microconidia, their shape and appearance were identified later. Fungal species of the genera*Aspergillus* and *Penicilium* were identified with the use of the keys by Raper and Fennell as well as Raper et. al., while the *Fusarium* species were identified using the key by Kwaśna (1991). Other species were identified based on the "Atlas of Clinical Fungi"(2010).

3. RESULTS. QUANTITATIVE AND QUALITATIVE RELATIONSHIPS BETWEEN INDOOR AND OUTDOOR MICROFLORA EXAMINED IN SUMMER AND SPRING

The studies were carried out in the summer of 2009 and in the spring of 2010, when the temperature of atmospheric air ranged between 15.2^{0}C and + 24.5^{0}C; the inside temperature in the poultry houses varied from 22^{0}C to 27^{0}C. Indoor relative air humidity was about 70-85 %, outdoor ca. 38-82%.

For both poultry houses, the indoor concentration of bacteria and moulds were always higher compared with the outdoor concentration at distance 10 m, 50 m and 100 m from the poultry houses. The number of microorganisms (as CFU/m^3) in the atmospheric air of both poultry houses ranged between 4 x 10^1– 7.2 x 10^3 for mesophilic bacteria, 0– 1.3 x 10^4 for staphylococci, 0 - 7 x 10^1 for coli group bacteria, and 2 x 10^1 – 1.3 x 10^4 for fungi (Fig. 2-5). *Salmonella sp.* were not found. On the other hand, the number of microorganisms inside both poultry houses was higher than in the surrounding area and ranged between 1.3 x 10^5– 5.2 x 10^5 for mesophilic bacteria, 1.4 x 10^5 – 2.6 x 10^5 for staphylococci, 2.0 x 10^2- 1.5 x 10^4 for coli group bacteria and 3.6 x 10^4 – 1,1 x 10^5 for fungi. *Salmonella sp.* similar like outdoor were not found.

Outside poultry house I mesophilic bacteria were the most numerous organisms in the spring and summer and formed about 55% of the local microbial community. Less numerous staphylococci and moulds constituted about 17% and 27.5%, respectively. The concentration of *Enterobacteriaceae* was fractional (0.5%). In contrast, outside poultry house II the most numerous group of bacteria were staphylococci (62%). Mesophilic bacteria and moulds constituted about 25% and 13% of the microbial community. However, air inside both poultry houses was characterized by a relatively small number of*Enterobacteriaceae.*

On the other hand, mesophilic bacteria were dominant and formed about 44% (poultry house I) and 56.5% (poultry house II) of the local microbial community, whereas staphylococci constituted 40% and 27% and moulds 16% and 7% for poultry house I and II. *Salmonella sp.* was not detected in neither poultry houses.

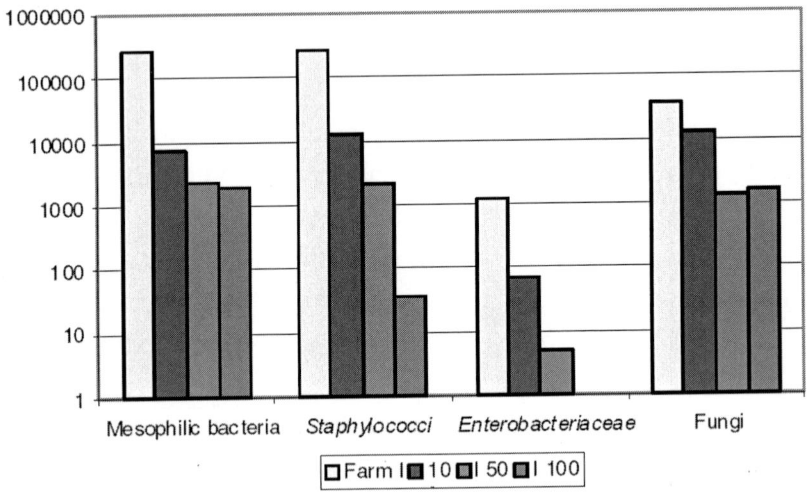

Figure 2. Number of microorganisms in the poultry house I in summer.

The level of outdoor air contamination was evaluated in the accordance with the Polish Norm (Table 1). Contaminations by mesophilic bacteria and moulds in areas surrounding poultry house I in the summer was the highest in sampling point I 10. In relation to the Polish Norms this site was heavily polluted. On the other hand, the number of staphylococci at all sampling sites in both poultry houses (except sampling points I 50 and II 100 in the spring) indicated a high contamination. In contrast, the air in the surrounding areas could be classified as medium-contaminated at all sites around the poultry houses II in the summer and in sampling point II 50 in the spring, with mesophilic bacteria. The evaluation based on the Polish Norm revealed that none of the researched measuring sites around the poultry house II was significantly contaminated by fungal microflora.

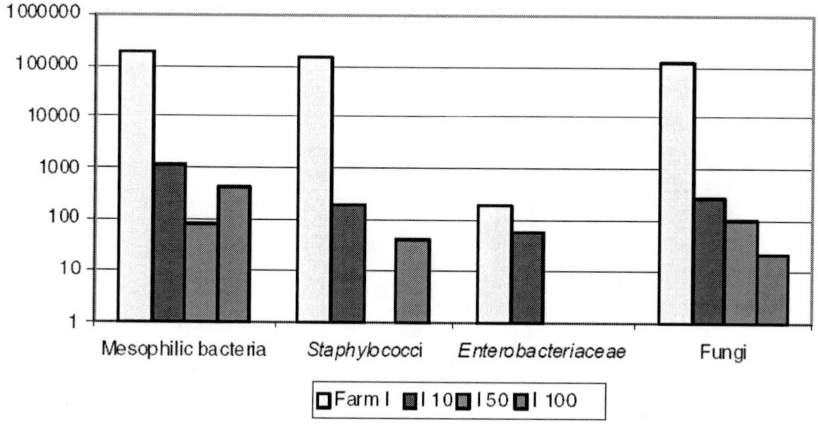

Figure 3. Number of microorganisms in the poultry house I in spring.

Fifteen species of bacteria detected in the indoor air represented 10 genera *Staphylococcus (S. aureus, S. xylosus, S. saprophitycus), Micrococcus (M. sedentarius, M. luteus), Enterococcus, Streptococcus (S. mitis), Corynebacterium (C. xerosis), Pseudomonas (P. aeruginosa, P. fluorescens), Citrobacter (C. farmerii), Escherichia (E. coli), Enterobacter (E. sakazakii, E. agglomerans)* and *Proteus (P. mirabilis)* – Table. 2. Quality outdoor set consisted of 19 species representing 10 bacteria genera: *Staphylococcus (S. sciuri, S. epidermidis, S. cohnii subsp., S. lentus,S. xylosus), Micrococcus (M. lylae, M. halobius, M. luteus), Streptococcus (S. mitis), Bacillus (B. mycoides, Bacillus sp.),*

Pseudomonas (P. aeruginosa, P. chlororaphis), Xantomonas (X. maltophila), Shigella (S. boydii), Providencia sp., Citrobacter (C. farmeri), Enterobacter (E. agglomerans) and *Proteus (P. mirabilis).* For the most part, the identified bacterial species were nonpathogenic or potentially pathogenic species, with the exception of *P. aeruginosa, S. aureus, Shigella sp., E. coli, P. mirabilis* which are regarded as primarily pathogenic.

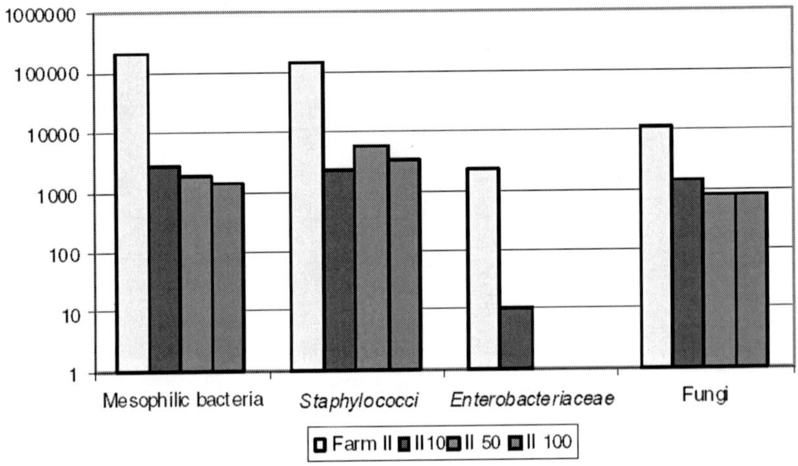

Figure 4. Number of microorganisms in the poultry house II in summer.

In this work we detected 30 species (16 in poultry houses and 23 in surrounding areas) representing 16 fungal genera: *Aspergillus (A. flavus* – Fig. 6), *A. niger, A terreus, A. nidulans, A. parasiticus, A. glaucus, A. clavatus* - Fig. 7), *Penicillium* – Fig. 10 *(P. chrysogenum, P. solitum, P. sticticus), Cladosporium (C. cladosporoides* – Fig. 8), *Alternaria (A. alternata* – Fig. 9, *A. tennuissima), Scopulariopsis (S. brevicaulis, S. acremonium), Fusarium* – Fig. 11 *(F. oxysporum, F. graminearum), Mucor (M.*

mucedo) Drechslera (D. graminae, Drechslera sp.), Verticillium sp., Mycelia sterilia,*Ulocladium sp., Trichoderma (T. viridae), Scedosporium sp., Candida (C. albicans, C. inconspicua, C. lambica), Rhodotorula (R. rubrum), Cryptococcus (C. laurentii)* – Table 3. The majority of these species are known as the potential respiratory allergens. The most common airborne moulds, both indoors and outdoors, were *Penicilium sp., Aspergillus sp., Cladosporium sp., Alternaria sp.,* and*Fusarium sp.* The yeast were sometimes the dominant fungus in the indoor air comparison with the outdoor air, where this group of fungi occur very sporadically.

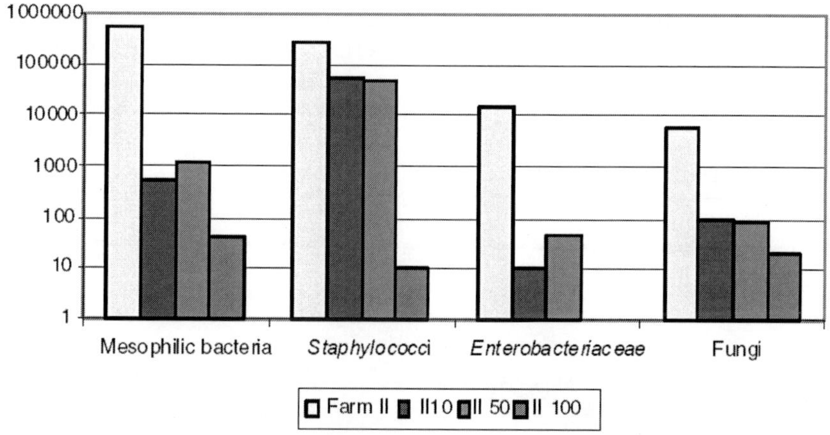

Figure 5. Number of microorganisms in the poultry house II in spring.

Among fungi identified from the farm I, there distinctly dominated the species belonged to genera*Candida spp.* which constituted on average, over 69% in summer and 82.5% in spring. In comparison with the predominated yeast inside poultry houses I, in outdoor air the most frequent species were moulds *A. flavus* (80% in sampling point I 50.36% in I 10 and 52.5% in I 100) and F.

oxysporum(26.9%, 13.5%, 16.2% in sampling point I 10, I 50, I 100, respectively). On the other hand, in the air surroundings poultry house II dominated species were *A. clavatus* (about 16%) and *A. alternata* (38.4%) in spring and *C. cladosporoides* (in summer) which constituted 68.5%. Whereas, *A. clavatus* (54.5%) and *A. flavus* (27.3%) were dominated species inside poultry house II.

Table 1. Polish Norm PN-89/Z-04111/02 and PN-89/Z-04111/03 (http://www.pkn.pl).

Polish Norm	Mesophilic bacteria	*Staphylococi*	Fungi
not pollution *	$< 1x10^3$	0	$3x10^3$-$5x10^3$
medium pollution **	$1x10^3$- $3x10^3$	<25	$5x10^3$-$1x10^4$
heavily pollution ***	$>3x10^3$	>25	$> 1x10^4$

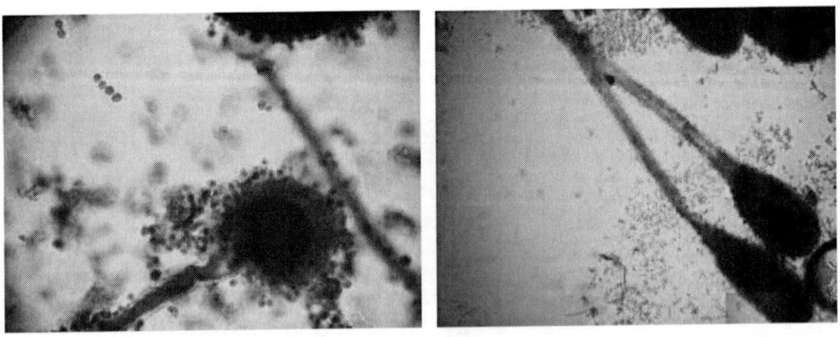

Figure 6, 7. Conidiophores of *A. flavus* and *A. clavatus*, oryg.

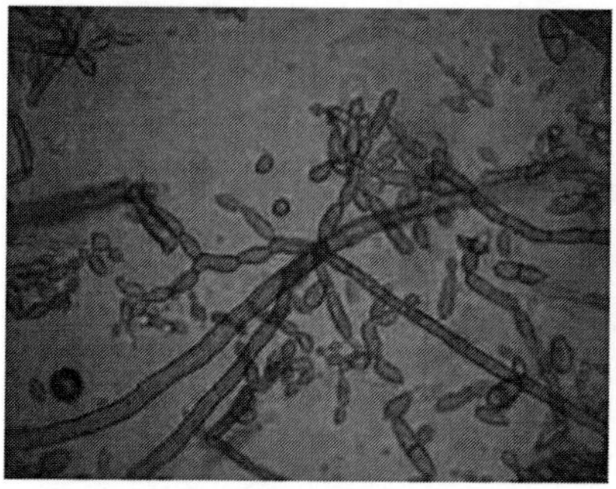 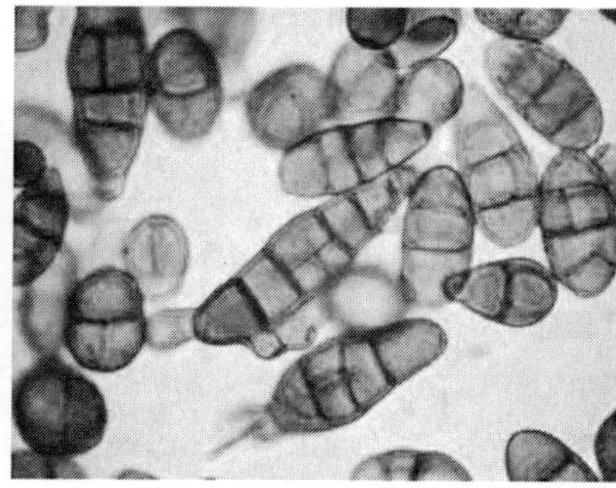

Figure 8, 9. Conidiophores of *C. cladosporoides* and poroconidia of *A. alternata*, oryg.

 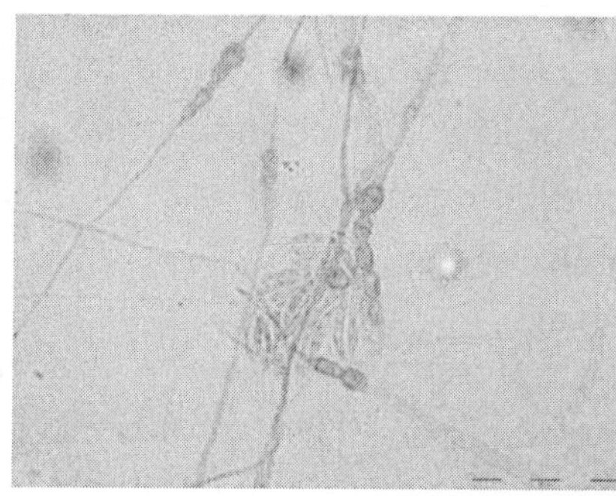

Figure 10, 11. Conidiophores of *Penicilium sp.* and macroconidia of *Fusarium sp.,* oryg.

4. DISCUSSION

The literature data usually show the air biopollutant concentration inside the poultry houses. According to many studies (Agranovski et al., 2007, Radon et al., 2002, Vučemilo et al., 2006, 2007) the number of bacteria in poultry houses ranged from 10^3 to 10^{10} CFU/m^3, and the concentrations of fungi was from 2.5 x 10^1 to 4.9

x 10^6 CFU/m^3. Much less is known about the relationships between the indoor and the outdoor biological pollution. In comparison with the rate of microorganisms contamination in the poultry houses, the concentration of bacterial and fungi in the air in surrounding areas was considerable lower and did not exceed the values found inside farms. Airborne bacterial and fungi levels measured in poultry houses I and II were always higher than in adjacent areas. Therefore we can suggested that the source of microorganisms are probably the farm objects. Baykov & Stoyanov 1999 also reported higher bacterial levels inside broiler farms than in nearby areas and the average values were similar to our results, i.e. value of mean of 16020 CFU/m3 for farmhouses and range of2060 CFU/m3to 386 CFU/m3 for immediate areas (10 m and 100 m from farm object, respectively). Very high concentration of microorganisms may reflect on insufficient ventilation in relation to the number of animals kept in poultry houses.

The presence of bacteria and fungi in poultry air is a natural phenomenon. Their primary source are the animals themselves, feed and litter. Microorganisms are a constituent of solid and liquid bioaerosols. This mostly refers to saprophytes, however pathogenic microorganisms were also found in the poultry houses air. Aerial count of pathogenic bacteria and fungi depends on the health condition of animals kept in the poultry houses. In addition, microorganisms count inside poultry farms air and monitoring of its emission from this building to the adjacent environment are important parameters for the assessment of the influence of poultry houses on the environmental pollution (Matković et al., 2006). In the present study, microbiological air contamination were determined at three sites at a distance of 10 m, 50 m and 100 m from the poultry houses. The results of total microorganisms count measurements outside the both poultry houses showed it to be lower than the total bacterial and moulds count inside the poultry houses. Number of microorganisms increase at 10 m distance from the poultry houses and gradually decreased to reach the lowest value at a distance of 100 m.

Table 2. Bacterial species isolated from the poultry houses I and II and from surroundings area during spring 2009 and summer 2010.

Genus	Species	Sampling site							
		Farm I	I 10	I 50	I 100	**Farm II**	II 10	II 50	II 100
Staphylococcus	*sciuri*		+				+		
	aureus					+	+	+	+
	epidermidis							+	
	cohnii subsp.2		+						
	lentus		+		+				
	xylosus	+	+						
	saprophitycus	+							
Micrococcus	*sedentarius*	+							
	lylae	+		+			+	+	
	halobius		+						
	luteus	+	+	+			+	+	
Enterococcus	*sp.*	+				+			
Streptococcus	*mitis*	+	+						
Bacillus	*mycoides*				+			+	
	sp.		+	+				+	
Corynebacterium	*xerosis*	+							
Psedomonas	*aeruginosa*	+	+		+				
	fluorescens	+							
	chlororaphis				+		+	+	

Xantomonas	*maltophila*				+				
Shigella	*boydii*		+						
Providencia	*sp.*		+	+					
Citrobacter	*farmerii*	+		+					
Escherichia	*coli*	+				+			
Enterobacter	*sakazakii*	+							
	agglomerans	+	+					+	
Proteus	*mirabilis*	+		+					

Table 3. Fungal species isolated from the poultry houses I and II and from surroundings area during study period (spring 2009 and summer 2010).

Genus	Species	Sampling site							
		Farm I	I 10	I 50	I 100	**Farm II**	II 10	II 50	II 100
Aspergillus	*flavus*	+	+	+	+	+			
	niger		+						
	terreus	+							
	glaucus		+						
	nidulans		+						
	parasiticus		+						
	clavatus					+	+	+	+
Penicilium	*solitum*	+							
	sticticus	+				+			
	chrysogenum		+		+				
Fusarium	*oxysporum*		+	+	+				
	graminearum		+	+	+				
Alternaria	*alternata*	+					+	+	+
	tennuissima					+		+	
Ulocladium	*sp.*				+				
Trichoderma	*viridae*	+	+						
Verticilium	*sp.*								+

Drechslera	*sp.*	+							
	gramineae				+				
Scopulariopsis	*brevicaulis*	+						+	+
	acremonium						+		
Cladosporium	*cladosporoides*					+	+	+	+
Mucor	*mucedo*	+	+			+	+	+	
Scedosporium	*sp.*					+	+	+	+
Mycelia	sterilia		+		+		+	+	
Rhodotorula	*rubrum*				+		+		
Candida	*albicans*	+				+			
	inconspicua	+							
	lambica							+	
Cryptococcus	*laurentii*	+				+			

Analyzing the results of the microbiological research about air pollution, it should be remembered that the results are temporary values, occurring at the moment the measurement. In connection with the physico-chemical properties of the air, the degree of contamination of the air can change diametrically within a few minutes (Donderski et al., 2005). Weather condition have a enormous influence on the count of microorganisms in the air. Temperature rise accompanied by rain scarcity can lead to a sudden increase in the concentration of microorganisms in the air. Consequently in summer with the weather conditions most friendly for the spread and development of numerous microorganisms, mesophilic bacteria, staphylococci, gram-negative bacteria and fungi were the more abundant in the air around the poultry houses than in early spring, where the temperature and humidity were lower.

Staphylococci seem to be a useful indicator bacteria (Schulz et al., 2004). Although this group of bacteria do not produce spores, they have the ability to survive in the air for a long time, which means spreading infections through the air. In the poultry houses I and II and outdoor air we observed very high numbers of potentially pathogenic staphylococci, what is really negative phenomenon. In our study the predominant species were coagulase negative. *S. aureus* and *S. saprophitycus* are pathogens for humans and the other species isolated in our studies may act as opportunistic pathogens in humans and animals.Karwowska 2005 also described the very high number of staphylococci in the air farming.

Moulds and yeast can live practically anywhere and have particularly favorable conditions inside the poultry houses. Among fungi recovered from the farm I, the species belonged to genera *Candida spp*. were dominated. This is a large group of potentially pathogenic species. They are usually an etiological factor in mycoses and less frequently, in mycoallergies. As they can produce toxins (e.g. candotoxins), the species can cause mycotoxycoses and increase microorganism sensitivity to some bacterial infections (Ejdys et al., 2009). On the other hand, in outdoor air the most frequent species were moulds *Aspergillus flavus* and *Fusarium oxysporum*. Amongst the several secondary metabolites produced by *A. flavus* are aflatoxins, the most toxic and potent carcinogenic natural compounds ever characterized. Whereas, the *F. oxysporum* is potentially producers of zearalenone, scirpentriol, NT-2 toxin, nivalenol, acetoxyscirpenediol, acetyl T-2toxin and others dangerous toxins (http://www.esgtesting. com/ Portal/Documents/Toxins%20and%20fungal%20origin.pdf).

In the air surroundings poultry house II dominated species were *Aspergillus clavatus, Alternaria alternata, Cladosporium cladosporoides*. Whereas, *A. clavatus* and *A. flavus* were the most frequent species inside poultry house II. We should emphasized that *A. alternata* is one of the most common fungi associated with asthma. Not only the presence of asthma but also persistence and severity asthma have been strongly associated with sensitization and exposure to *A. alternata* (Salo et al., 2006). On the other hand, the presence of opportunistic pathogens from the genus Aspergillus poses a risk of invasive aspergillosis in farm workers and those living in the proximity of the farms. According to the data obtained by other authors (Soliman et al., 2009) fungi e.g. *Candida albicans, Aspergillus niger, A. nidulans, Penicilium sp.* and *Mucor sp.* were prevalent in broiler farms in Egypt. However, Romanowska-Słomka and Mirosławski 2009 described the occurrence of the moulds and yeast*Aspergillus sp., Penicillium sp., Candida sp. and Cryptococcus sp.* in poultry houses. Other investigators (Agranovski et al. 2007) isolated and identified many fungal strain, including genera:*Cladosporium, Aspergillus, Penicillium, Scopulariopsis, Fusarium, Epicoccum, Mucor, Trichophyton, Alternaria, Ulocladium, Basidiospores, Acremonium, Aureobasidium, Drechslera, Pithomyces, Chrysosporium, Geomyces and Rhizomucor* from farming areas. The presence of such fungi in farmhouses was proved by the results of this study.

The highest total *Enterobacteriacea* counts were found in indoor air and in areas nearby poultry houses I and II. In this present studies both in farms and surroundings areas *Escherichia coli, Proteus mirabilis, Shigella boydii, Citrobacter farmerii, Enterobacter agglomerans, E. sakazakii, Klebsiella pneumoniae, Providencia sp.* were identified. Different results was observed by Vučemilo et. al who found four dominating species of the *Enterobacteriaceae* family: *E. coli, Pantoea sp., Serratia plymuthica* and*Serratia marcescens*. According to Lues et. al 2007 *E. coli* and the other members of the coliforms bacteria could be a good indicators of air contamination.

5. CONCLUSIONS

The farming buildings are emitters of the considerable amounts of microbiological contaminants into the atmospheric air. This high emission of potentially pathogenic microorganisms via aerosols from animal housing facilities to the outdoor environment may constitute a considerable risk to human health and environmental pollution.

So far, in literature there are no reliable data about relationships between the indoor and the outdoor biological pollution. This study contributes to the understanding of the level concentration of bioaerosol and its composition with regard to the different distance from farms. The quantity and quality of microbial analysis shows both different bacteria genera and fungi in indoor and outdoor microbiological contamination. Comparing our own results with available literature data on the indoor and outdoor air biopollutant concentration in the poultry houses enables to understand the distribution process.

REFERENCES

1. V. Agranovski, T. Reponen, Z. Ristovski, 2007Survey of bioaerosol emissions from Australian poultry buildings. European Aerosol Conference. Salzburg, Abstract, 28.

2. R. Araujo, J. P. Cabral, 2010Fungal air quality in medical protected environments. In: Air Quality, Ashok Kumar, 357 382Sciyo, 978-9-53307-131-2Publisher: Sciyo.

3. B. Bakutis, E. Monsteviliene, G. Januskeviciene, 2004Analyses of airborne contamination with bacteria, endotoxins and dust in livestock barns and poultry houses. Acta Veterinaria Brno, 73 283 289

4. B. Baykov, M. Stoyanov, 1999Microbial air pollution caused by intensive broiler chicken breeding. Federation of European Microbiological Societies. Microbiology ecology, 29 389 392

5. B. Crook, A. Easterbrook, S. Stagg, 2008Exposure to dust and bioaerosols in poultry farming. Summary of observations and data. Prepared by the Health and Safety Laboratory for the Health and Safety Executive.

6. S. Dehoog, J. Guarro, J. Gene, M. J. Figureas, 2008Atlas of Clinical Fungi. Centraalbureall voor Schiielcultures/ Universitat Rovira i Virgili.

7. W. Donderski, M. Walczak, M. Pietrzak, 2005Microbiological contamination of fair within the city of Toruń. Polish Journal of Environmental Studies, 14 2 223 230

8. J. Dutkiewicz, 1987Bacteria in farming environment. European Journal of Respiratory diseases, 71 154 71 88

9. E. Ejdys, J. Michalak, K. M. Szewczyk, 2009Yeast-like fungi isolated from indosr air in school buildings and the surrounding outdoor air. Acta Mycologica, 44 1 97 107

10. E. Lonc, K. Plewa, 2009Microbiological air contamination in poultry houses. Polish Journal of Environmental Studies, 14 4 445 449

11. E. Karwowska, 2005Microbiological air contamination in farming environment.. Polish Journal of Environmental Studies, 19 1 15 19

12. H. Kwaśna, J. Chełkowski, P. Zajkowski, 1991Grzyby (Mycota). Tom XXII. PAN, Warszawa Kraków, 136 ss.

13. J. Lues, M. Theron, P. Venter, H. Rasephei, 2007Microbial composition in bioaerosols of a high-throughput chicken-slaughtering facility. Poultry Science, 86 142 149

14. A. Pasqualotto, 2008Differences in pathogenicity and clinical syndromes due to Aspergillus fumigatus and Aspergillus flavus. Medical mycology, 1 10

15. Polish Norm (PN-89/Z-04111/02). Air purity protection. Microbiological testing.Determination number of the bacteria in the atmospheric air (emission) with sampling by aspiration and sedimentation methods.

16. Polish Norm (PN-89/Z-04111/03). Air purity protection. Microbiological testing.Determination number of the fungi in the atmospheric air (emission) with sampling by aspiration and sedimentation methods.

17. K. Matković, M. Vučemilo, B. Vinković, Ž. Pavičić, S. Matković, M. Benić, 2009Airborne fungi in a dairy barn with emphasis on microclimate and emission. Veterinarski Arhiv, 79 3 207 218

18. K. Matković, M. Vučemilo, B. Vinković, B. Šeol, Ž. Pavičić, S. Matković, A. Tofant, S. Matković, 2006Effect of microclimate on bacterial count and airborne emission from dairy barns on the environment. Annals of Agricultural and Environmental Medicine, 13 349 354

19. A. Nevalainen, 2007Bio-aerosols as exposure agents in indoor environments in relation to asthma and allergy. National Public Health Institute, Department of Environmental Health POB 95, FI-70701 Kuopio, Finland.

20. D. Pomorska, L. Larsson, C. Skórska, J. Sitkowska, J. Dutkiewicz, 2009Levels of bacterial endotoxins in the samples

of settled dust collected in animal houses. Bulletin of the Veterinary Institute in Pulawy, 53 37 41

21. K. Radon, B. Danuser, M. Iversen, O. Monso, Ch. Weber, J. Hartung, U. Palmgren, D. Nowak, 2002Air contaminants in different European farming environments. Annals of Agricultural and Environmental Medicine, 9 41 48

22. I. Romanowska-Słomka, J. Mirosławski, 2009Biological hazards in an industrial poultry farm- research results. Bezpieczeństwo Pracy, 7 16 19

23. K. B. Raper, Ch. Thom, D. I. Fennell, 1949A manual of the Penicillia. The Williams & Wilkins Company, Baltimore, USA.

24. K. B. Raper, D. I. Fennell, 1965The genus Aspergillus. The Williams & Wilkins Company, Baltimore, USA.

25. P. Salo, S. Arbes, M. Sever, R. Jaramillo, R. Cohn, S. London, D. Zeldin, 2006Exposure to Alternaria alternata in US homes is associated with asthma symptoms. Journal of Allergy and Clinical Immunology, 118 4 892 898

26. R. Schierl, A. Heise, U. Egger, F. Schneider, F. Eichelser, S. Neser, D. Nowak, 2007Endotoxin concentration in modern animal houses in southern Bavaria. Annals of Agricultural and Environmental Medicine, 14 1 129 136

27. J. Schulz, J. Hartung, L. Seedorf, C. Formosa, 2004Staphylococci as an indicator for bacterial emissions from a broiler house. International Soeciety for Animal Hygiene-Sant-Maolo, 75 78

28. S. Soliman, M. Sobeih, M. Hussein, H. Abdel-Latiff, A. Moneim, 2009Seasonal epidemiological surveillance on

bacterial and fungal pathogens in broiler farms in Egypt. International Journal of Poultry Science, 8 8 720 727

29. M. Siemińki, 2001 Environmental health threats, Wydawnicto Naukowe PWN, 8-30113-561-1

30. P. Srikanth, S. Suchithra, R. Steinberg, 2008Bio-aerosols in indoor environment: composition, health effects and analysis. Indian Journal of Medical Microbiology, 26 4 302 312

31. M. Vučemilo, B. Vinković, K. Matković, 2006Influence of broiler age on airborne pollutant content in poultry house. Krmiva, 48 3 6

32. M. Vučemilo, K. Matković, B. Vinković, S. Jakšić, K. Granić, N. Mas, 2007The effect of animal age on air pollutant concentration in a broiler house. Czech Journal of Animal Science. 52 6 170

CHAPTER 2

Mortality and Pathology Associated with Highly Pathogenic Avian Influenza H5N1 Outbreaks in Commercial Poultry Production Systems in Nigeria

Olatunde Babatunde Akanbi[1,2] and Victor Olusegun Taiwo[2]

[1]Central Diagnostic Laboratory, National Veterinary Research Institute, PMB 1, Vom 930001, Plateau State, Nigeria

[2]Department of Veterinary Pathology, Faculty of Veterinary Medicine, University of Ibadan, Ibadan 200284, Oyo State, Nigeria

ABSTRACT

Commercial layer-type, pullet, cockerel, and broiler chicken flocks infected with highly pathogenic avian influenza (HPAI) H5N1 in Nigeria between 2006 and 2008 were investigated for morbidity, mortality, and pathology. Of the one hundred and fifty-three (153) farms confirmed with HPAI infection, one hundred and twenty-seven (127) were layer-type farms, nine (9) were pullet and broiler farms each, and eight (8) were cockerel rearing farms. This study revealed the morbidity and mortality of a total of 939,620 commercial layer chickens, 16,421 pullets, 3,109 cockerels, and 6,433 broilers. Mortality rates were 11.11% in commercial layers, 26.84% in pullets, 45.51% in cockerels, and 73.92% in broilers in a total of eighteen (18) states and the Federal Capital Territory, Abuja. A total of 316 carcasses were examined of which 248 were commercial layer, 25 were pullet, 14 were cockerel, and 29 were broiler. Main clinical and pathologic findings were observed in the nervous, circulatory, respiratory, integumentary, musculoskeletal, hemopoietic, gastrointestinal, and reproductive systems and, occasionally, lesions were generally nonspecific and multisystemic. Lesions occurred more frequently, severely, and in most of the carcasses examined, irrespective of chicken type.

1. INTRODUCTION

The index HPAI-H5N1 was confirmed in Nigeria at a commercial poultry farm in Kaduna state, and, by the end of the initial outbreak, over 46,000 poultry [1, 2] had been destroyed. This outbreak brought the Asian strain of highly pathogenic avian influenza (HPAI) H5N1 into Africa for the first time in the beginning of January, 2006 [3–7] in Nigeria. Nigeria lost 945,862 birds of various species at mid-January [8], 2007. Since its emergence in Africa in 2006, avian influenza viruses of the H5N1 subtype have spread rapidly to poultry farms in several African countries. It has caused deaths of millions of birds in Africa [9], as is the case in other affected parts of the world. Highly pathogenic avian influenza (HPAI) is caused by AIVs that are extremely virulent [10], causing up to 100% mortality in domestic chickens. The H5N1 highly pathogenic avian influenza (HPAI) virus has been a great concern not only for the poultry industry but also for human health [11] since 1997. Following the index HPAI-H5N1 in

Nigeria, several efforts have been directed to increased surveillance and diagnosis of the disease. This has led to increased diagnostic capacity, reporting, and research into the diverse genotype of H5N1 isolates from Nigeria. The commercial poultry production system in Nigeria as categorized by the FAO [12], which stems from the two major systems: rural poultry production and commercial poultry production, was affected by the HPAI outbreaks. Also, several attempts were made to determine the losses in terms of the number of poultry, monetary, and social values as a result of the disease burden [8] in Nigeria. These attempts have been at some point in the course of the disease [8] or aimed at a particular region of the country [13] due to some identified constraints. Also no overall, detailed clinical and pathological record has been given on a case by case basis for all the outbreaks as seen from different states and regions of the countries which could highlight factors involved in the spread of the disease. In early 2006, Joannis et al. 2006 [4], in their letter to the Veterinary Record, mentioned a few lesions seen in the index case. In addition, Adene et al. 2006 [2] described the clinicopathological and husbandry features associated with the maiden diagnosis of AI in Nigeria. Other reports were limited to cases seen in 2006 [14, 15] only. Akanbi et al. 2007 [5] had earlier reported lesions which depict circulatory disturbances, respiratory, intestinal, and reproductive system pathology in selected cases. This study aimed at achieving the determination of the losses on a state by state basis and the total losses recorded during the entire course of the disease in Nigeria and also aimed at highlighting the morbidity, mortality, and pathology in the commercial poultry production systems as practiced in Nigeria.

2. MATERIALS AND METHODS

Data. All the data including state, location, farm flock size, and morbidity and mortality records used in this study were supplied directly by clients who reported and submitted carcasses of layers, pullet, cockerels, and broilers for postmortem examination and avian influenza diagnosis at the National Veterinary Research Institute (NVRI), Central Diagnostic Laboratory, Vom.

Carcasses. A total of three hundred and sixteen (316) carcasses from one hundred and fifty-three (153) farms from eighteen (18) states and the Federal Capital Territory, Abuja, were examined. These carcasses were submitted directly by clients for postmortem examination and avian influenza diagnosis and were selected based on laboratory confirmation of HPAI H5N1 virus infection at the NVRI laboratory.

Pathology. Carcasses of the commercial birds that died after natural infection with HPAI were submitted for pathological examination. Following postmortem examination of the carcasses, sections of liver, heart, spleen, kidney, lung, trachea, proventriculus, gizzard, duodenum, ileum, cecum, and brain were removed and fixed in 10% buffered formalin. All tissue samples were then embedded in paraffin, sectioned at 5 μm, mounted on clean glass slides, and stained with hematoxylin and eosin (H&E) stains for histopathologic examination using low and high powered field of Carl Zeiss or Nikon binocular microscope.

Virus Isolation. For virus isolation, samples of liver, heart, spleen, kidney, lung, trachea, duodenum, and caecum were removed and submitted for virological investigation, where the homogenates of the trachea, lung, liver, spleen, and kidney were inoculated via the allantoic cavity into 9–11-day-old embryonating chicken eggs [16] obtained from specific antibody negative chickens.

3. RESULTS

3.1. Data Set Distribution

During the period under review (2006–2008), a total of eighteen (18) states (Adamawa, Bauchi, Bornu, Jigawa, Kaduna, Kano, Katsina, Sokoto, Plateau, Nasarawa, Kwara, Lagos, Ogun, Oyo, Rivers, Edo, Anambra, and Enugu) and the Federal Capital Territory, Abuja, had confirmed HPAI H5N1 outbreaks in commercial layer, pullet, cockerel, and broiler (Table 1). A total of one hundred and fourteen thousand, nine hundred and twenty-nine (114,929) commercial chickens died from a total flock size of nine hundred and sixty-five thousand, five hundred and eighty-three (965,583) before the rest of the flocks were stamped out. The north-western region recorded the highest number of mortalities in commercially raised chickens as a result of natural infection with HPAI in Nigeria. The figure (69,741) represents 60.68% of the total number (114,929) of commercially raised chickens that died as a result of HPAI (Table 1). While the total number of chicken losses as a result of HPAI death and stamping out in this region (438,386) was second to the south-western region where commercial bird losses totaled 445,320 and the number that died as a result of HPAI when the reporting was done was 26,878, being 23.38% of the total number of commercial chickens that died as a result of HPAI.

3.2. Mortality

Mortality rates were the highest in broilers flocks (73.92%) and the least in layers flocks (11.11%). The mortality rate was 45.51% in cockerels and was 26.84% in pullet. A breakdown of the figures, which showed layer flocks, has been the most affected commercial chicken type with a total bird loss of 939,620, representing 97.31%

of the commercially raised chickens. Of the 939,620 layer population affected in 127 farms, 104,351 layers, representing 90.79% of 114,929, died as a result of direct HPAI infection in layer flocks. Eight hundred and fifty thousand, six hundred and fifty-four (850,654) birds, which represent the difference between the flock size and the number of the dead, were stamped out. Broilers were the second most affected commercial chicken type of which 6,433, representing 0.66% of the total flock size (965,583) of commercially raised chickens, were affected by HPAI. Of the 6,433 broilers affected, 4,755 broilers, representing 4.14% of the total number of commercially raised chickens (114,929), died as a result of direct HPAI infection, while the remaining broilers were stamped out. Pullets were the next affected commercial chicken type of which 16,421 (11.70%) were affected and 4408 (3.84%) were reported to be dead. Cockerel was the least affected commercial chicken type of which 3109 (0.32%) were affected and 1,415 (1.23%) were reported to be dead.

3.3. Clinical Signs and Pathology

Main clinical and pathologic findings were observed in the nervous, circulatory, respiratory, integumentary, musculoskeletal, gastrointestinal, and reproductive systems and occasionally lesions are multisystemic. Signs observed and reported are sudden death, high mortality, weakness, and recumbency. Others ranged from nasal discharges, dyspnea, coughing, sneezing, diarrhea, shank hyperemia and haemorrhage, inability to stand, ataxia, and torticollis. In layers, egg structural abnormalities such as shell-less egg, white-colored eggs, and soft eggs were reported. Lesions observed in the circulatory system included congestion (Figures 1 and 2) and cyanosis of comb and wattle, comb and wattle edema, and facial and subcutaneous edema. Within the respiratory system, there were airsacculitis and pneumonia. There was petechiation to ecchymoses of the proventricular (Figure 3) and intestinal mucosa with resultant enteritis in the gastrointestinal system. Integumentary system lesions are mainly cyanosis, edema, and ecchymotic hemorrhages while there were inflammatory, degenerative, and necrotic lesions in the musculoskeletal system. In adult birds, mainly layers reproductive lesions were observed and they were mainly ovarian follicular ecchymotic hemorrhages (Figure4).

Layers. Most of the commercially raised layers (67%) exhibited lesions of the circulatory system, mainly cyanosis of comb and wattle with occasional facial edema. Only 20% showed nervous signs and brain lesions of neuronal and Purkinje cell necrosis of cerebrum and cerebellum, respectively. Respiratory lesion of airsacculitis and pneumonia (Figure 4) with vascular congestion (Figure 5) was evident in 43.5% with more than half having pneumonia. Of the 70 clinical reports of diarrhea, only 42 had enteritis and 18 were hemorrhagic. Enteric petechiation and ecchymoses were observed in 45 of the carcasses while pancreatic necrosis was occasionally encountered (Figures 6 and 7). Splenic necrosis and loss of lymphoid cells were also seen in the carcasses (Figure 8). A few carcasses (6.4%) showed integumentary system lesions; these are mainly cyanosis, edema, and ecchymotic hemorrhages. Only 13.7% had muscular hemorrhages with necrosis, and/or myositis. Reproductive lesions were only observed in these layers, and these were observed in 15.3% only. These were mainly ovarian follicular ecchymotic hemorrhages and structure abnormality. Forty-one (41) percent of the layers had multisystemic lesions. Shank hyperemia (Figure 9) and haemorrhage were also frequently seen.

Table 1. Commercial poultry losses in HPAI infection in Nigerian states and region.

S/N	Zone	State	Flock size	Number of the dead	Chicken type
		Adamawa	55	36	Cockerel
	Northeast	Bauchi	28731	4637	Layer, pullet, broiler
		Bornu	800	207	Layer, pullet, cockerel
	Regional subtotal		**29586**	**4880**	
		Jigawa	13786	3000	Layer
		Kaduna	87435	51333	Layer, pullet, broiler, cockerel
	Northwest	Kano	234998	7930	Layer, cockerel
		Katsina	95605	6778	Layer
		Sokoto	6562	700	Layer
	Regional subtotal		**438386**	**69741**	
		Abuja	236	165	Layer, broiler
	North central	Plateau	24210	4712	Layer, pullet, broiler, chick
		Nasarawa	58	56	Broiler, cockerel, mixed
		Kwara	6685	1485	Layer
	Regional subtotal		**31189**	**6418**	
		Lagos	291780	21632	Layer, pullet, broiler
	Southwest	Ogun	138536	642	Layer, cockerel
		Oyo	15004	4604	Layer, broiler, cockerel
	Regional subtotal		**445320**	**26878**	
	South-south	Rivers	1200	200	Layer
		Edo	6500	5109	Layer, pullet, broiler
	Regional subtotal		**7700**	**5309**	
	Southeast	Anambra	13352	1653	Layer
		Enugu	50	50	Broiler
	Regional subtotal		**13402**	**1703**	
	National total		**965583**	**114929**	

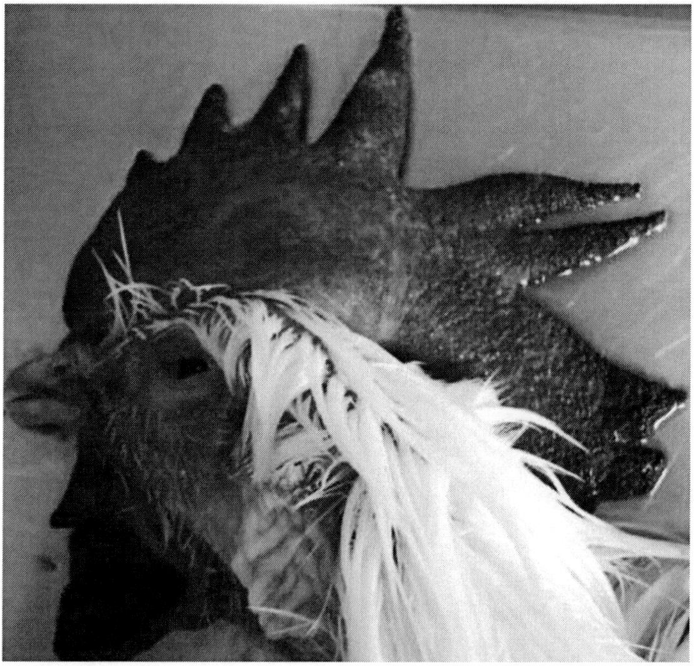

Figure 1. Layer chicken, head showing swollen comb (arrow) and congestion (arrow head).

Pullet. Of the twenty-five (25) pullet carcasses examined, 60% had report of the circulatory system signs, mainly cyanosis of comb and wattle with occasional facial edema. Only 4% showed nervous lesion in the brain such as neuronal and Purkinje cell necrosis of cerebrum and cerebellum, respectively. Respiratory lesions of nasal exudation, airsacculitis, and pneumonia were evident in 32%. Enteric lesion was observed in 3 (12%) of the 25 carcasses examined. Twenty-eight (28) percent had muscular and shank hemorrhages with necrosis, and/or myositis. Twenty-four (24) percent of the pullets which died suddenly were from flocks with high mortality and also had multisystemic lesions. No lesion was observed in the integumentary and reproductive systems, respectively, in all carcasses examined.

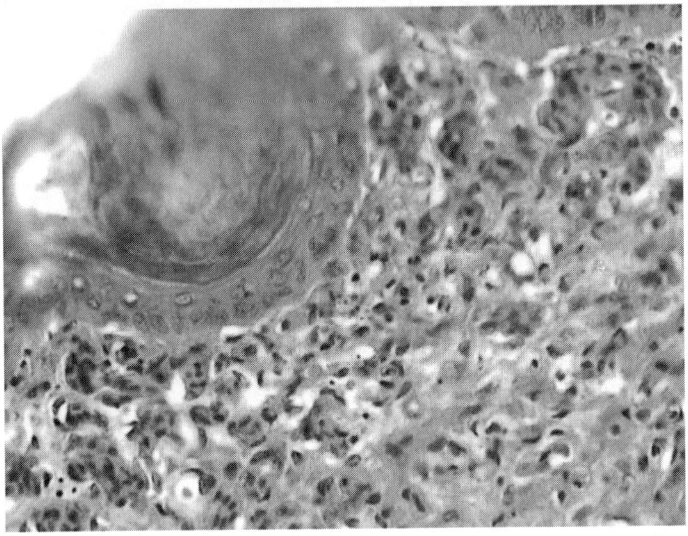

Figure 2. Layer chicken, comb showing severe dermal vascular congestion (arrow heads) H&E, ×400.

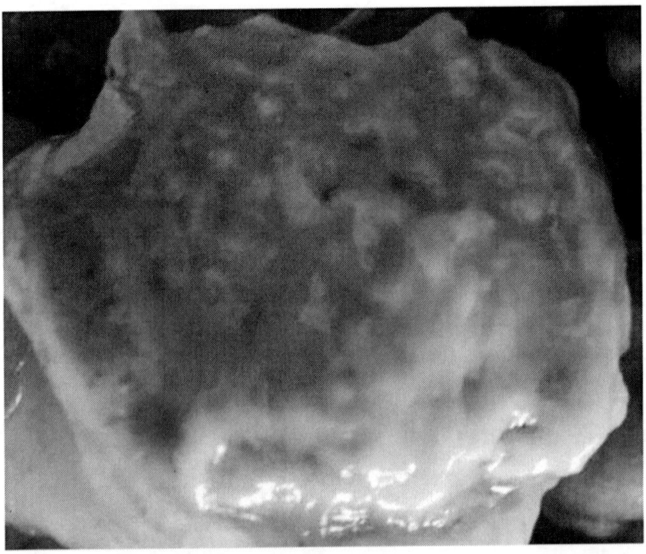

Figure 3. Layer chicken, proventriculus showing swollen glands with excess mucous exudates (arrow) and mild multifocal to diffuse petechiae hemorrhages (arrow head).

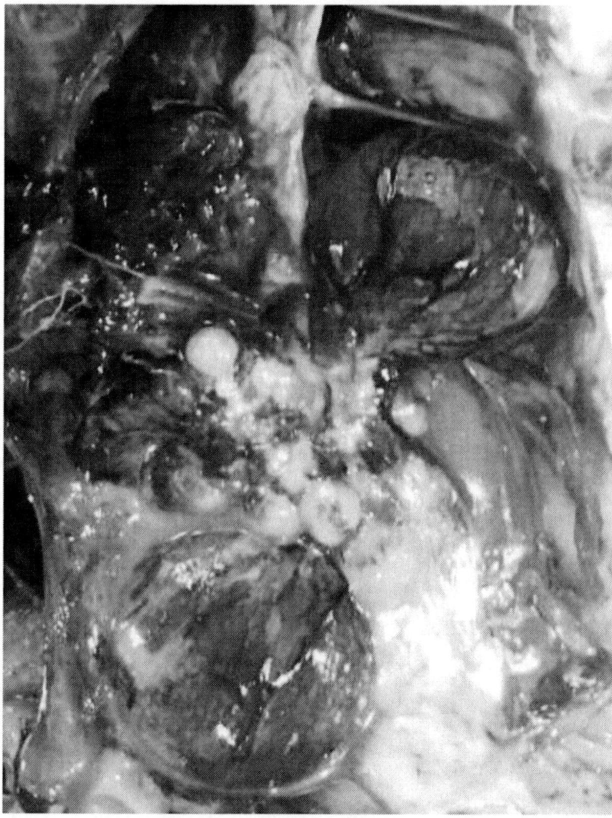

Figure 4. Layer chicken, thoracic and abdominal cage, showing congested and consolidated lungs (arrow) and haemorrhagic ovarian follicles (arrow head).

Broiler. Of the twenty-nine (29) broiler carcasses examined, 82.7% had lesions of the circulatory system, mainly cyanosis of comb and wattle with occasional facial edema. Only 4 (13.7%) showed brain lesions of neuronal and Purkinje cell necrosis of cerebrum and cerebellum, respectively. Respiratory lesions of airsacculitis and pneumonia were evident in fourteen (48.2%) carcasses. Enteric lesion was observed in 9 (31%) of the 29 carcasses examined. Five (17.2%) carcasses had muscular and hemorrhages with necrosis and/or myositis. Ten (34.4%) of the broilers which died suddenly were from flocks with high mortality and had multisystemic lesions. No lesion was observed in the integumentary and reproductive systems, respectively, in all carcasses examined.

Cockerel. Of the fourteen (14) cockerel carcasses examined, 64% had lesions of the circulatory system, mainly cyanosis of comb and wattle with occasional facial edema. Only 5 (35.7%) showed nervous lesions of neuronal and Purkinje cell necrosis of cerebrum and cerebellum, respectively. Respiratory lesions of airsacculitis and pneumonia were evident in five (35.7%) carcasses. Enteric lesion was observed in 6 (42.8%) of the 14 carcasses examined. Eight (57.1%) carcasses had muscular and shank hemorrhages with necrosis and/or myositis. Five (35.7%) of the pullets which died suddenly were from flocks with high mortality and had multisystemic lesions. No lesion was observed in the integumentary and reproductive systems in all carcasses examined.

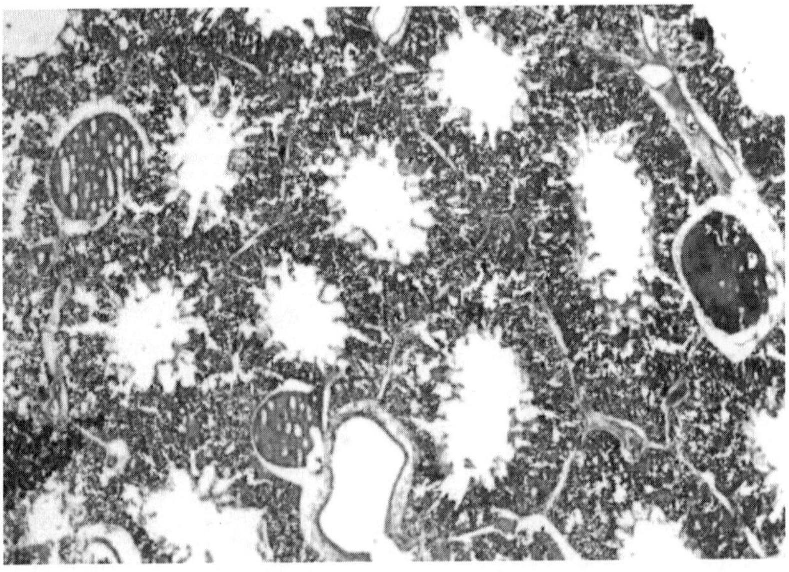

Figure 5. Layer chicken, lung showing severe multifocal vascular congestion (arrow), H&E, ×400.

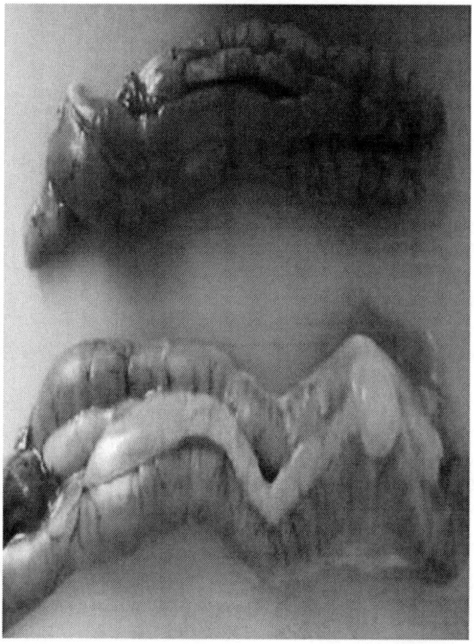

Figure 6. Layer chicken, duodenal loop showing multifocal necrosis on the pancreas (arrow heads).

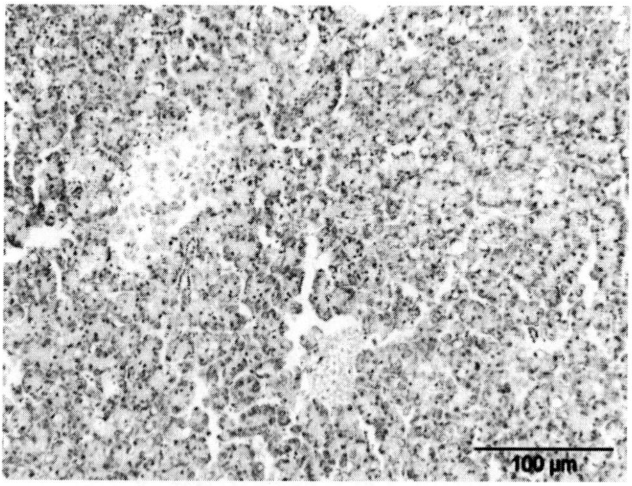

Figure 7. Layer chicken, pancreas showing diffuse necrosis of acinar cells of the pancreas (arrow heads) H&E.

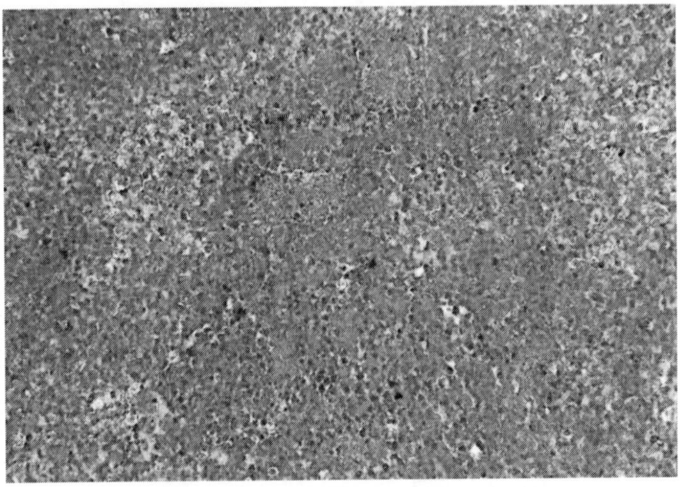

Figure 8. Layer chicken, spleen showing severe diffuse lymphoid necrosis and depletion (arrow heads) H&E, ×400.

Figure 9. Layer chicken, shank showing hyperemia.

4. DISCUSSION

The north-western region recorded the highest number of mortalities in commercially raised chickens as a result of natural infection with HPAI in Nigeria. This represents 60.68% (69,741) of the total number of chickens which died as a direct result of HPAI infection. The total number of chickens which died as a direct result of HPAI infection and those that were stamped-out in this region was 438,386, only second to the total number of chickens which died as a direct result of HPAI infection and those that were stamped out in the south-western region which stood at 445,320. Indeed, south-western Nigeria, particularly the states surrounding the city of Lagos [17], holds much of Nigeria's poultry industry. It is estimated that over 65% of Nigeria's commercial poultry are located in the five southern states [17] of Lagos, Ogun, Oyo, Osun, and Ondo. A very high number of poultry deaths and losses were also reported in a north-western regional [13] avian influenza investigation in the north-western region, apart from the fact that it was in this region where the first outbreak of HPAI in Nigeria was reported [4, 18] in a farm that had predominantly commercial birds [2, 14]. Although the total poultry losses were higher in the southwest region, the total number of chickens which died as a direct result of HPAI infection was 26,878 being 23.38% of the national total number of commercially raised chickens that died as a direct result of HPAI infection, as compared with 60.68% (69,741) in the northwest. This difference may be as a result of the fact that the farms in the southwest are more organized and densely stocked as compared with those in the northwest which are not well organized and not as densely stocked. The finding of high mortality in the north-western states is consistent with earlier reports [2, 13]. In the north-western regional analysis [13] which excluded bird losses in Sokoto state, four hundred and eighty thousand, three hundred and seventy-eight (480,378) birds were reported as being lost as a result of HPAI incursions and stamping out. The deviation of this finding from ours may be as a result of nonseparation between the various sectors of poultry production systems, as it includes backyard flocks and noncommercial and commercial flocks. The north-central states had a mortality of 6418 (5.58%) being the total number of commercial chickens that died as a result of HPAI when the reporting was done, while the total number of chicken losses as a result of natural death and stamping out policy in this region was 31,189. The south-south region lost more birds to direct HPAI infection (5309) as compared to the north-eastern region (4880) but more birds (29586) were destroyed in the northeast than in the south-south where 7,700 birds were destroyed. The southeast lost the least number of birds to HPAI (1703) but had a high number of birds' losses to HPAI and stamping out (13402). Under the commercial poultry production sector practiced in Nigeria [17], the commercial layer-type chicken was the most hit by the highly pathogenic avian influenza outbreaks experienced by the Nigerian poultry industry. With a total flock size of 939,620 lost in 127 farms having confirmed outbreaks, the commercial layer-type chicken suffered a major economic loss and devastation. This is attributed to the large number of laying birds in individual flocks as compared to other chicken types under these sectors. This significantly affected the economy not only because a whooping sum of N631 million (US$5.43 million) was paid as compensations to farmers [19] but also because laying hens contribute huge resources to the national poultry flock and this emphasizes the importance of commercial layer flocks [20] for Nigeria's economy. It was also observed that the average mortality rate was the least in commercial layers (11.11%) and the highest in broilers (73.92). This may be as a result of the cage housing of commercial layers which restrict movement and contact between HPAI infected ones and noninfected ones compared with broilers raised in deep litter and which also share water and feed sources. Average mortality rate was higher in cockerels (males = 45.51%) than in pullets (females = 26.84%). The necropsy finding observed in the 316 carcasses examined of which 248 were commercial layer, 25 were pullet, 14 were cockerel, and 29 were broilers cut across more than one system and occurred more frequently and severely and in most of the carcasses examined irrespective of chicken type. In commercial layers, the systems affected included mainly circulatory, respiratory, and intestinal systems as observed by earlier investigators from the index case of HPAI [2, 4] in Nigeria. Only 20% of the 248 layer carcasses had neurologic signs and lesions contrary to what was observed in the index case where only young stocks had nervous system signs [2]. Also, other investigators reported presence of nervous signs and lesion with no particular reference to age [4] and with reference to adult birds [15]. Eighty-two percent (82.7%) of the broilers showed signs and lesions of circulatory disturbances followed by 67% in layers. More broilers and cockerels showed neurological signs and lesion more than in layers. This work was able to document the total HPAI bird losses in commercially raised chickens in Nigeria on a state by state cases and region by region cases, previously unreported. It also showed that mortality was the highest in broiler and males (cockerels) had more death than females (pullets).

5. CONCLUSION

This study was able to document the losses in HPAI infected commercial chickens in Nigeria and the overall pathological findings of HPAI infection in chickens in Nigeria previously unreported.

CONFLICT OF INTERESTS

The authors declare that there is no conflict of interests regarding the publication of this paper.

ACKNOWLEDGMENTS

The authors would like to acknowledge the technical assistance of Ben Bako, Rebecca Kaye, Alesa Mada, and Victoria Olabode and the cooperation of the National Veterinary Research Institute, Vom, Nigeria.

REFERENCES

1. "N.V.R.I records," National Veterinary Research Institute, Case database, Vom, Nigeria, 2006.

2. D. F. Adene, A. M. Wakawa, P. A. Abdu, et al., "Clinico-pathological and husbandry features associated with the maiden diagnosis of Avian Influenza in Nigeria," Nigeria Veterinary Journal, vol. 27, no. 1, pp. 32–38, 2006.

3. M. Enserink, "Avian influenza. H5N1 moves into Africa, European Union, deepening global crisis.,"Science., vol. 311, no. 5763, p. 932, 2006.

4. T. Joannis, L. H. Lombin, P. de Benedictis, G. Cattoli, and I. Capua, "Confirmation of H5N1 avian influenza in Africa," Veterinary Record, vol. 158, no. 9, pp. 309–310, 2006.

5. B. O. Akanbi, J. P. Teifke, P. S. Ekong, E. T. Ighodalo, A. Ogunsan E, and A. A. Makinde, “The epidemiology, pathology and immunohistochemistry of Nigerian highly pathogenic avian influenza (HPAI) H5N1 virus infection in chickens,” in Proceedings of the 44th Annual Congress of the Nigerian Veterinary Medical Association (NVMA) Conference, pp. 162–165, Effurun, Nigeria, October 2007.

6. M. F. Ducatez, C. M. Olinger, A. A. Owoade et al., “Molecular and antigenic evolution and geographical spread of H5N1 highly pathogenic avian influenza viruses in western Africa,” Journal of General Virology, vol. 88, no. 8, pp. 2297–2306, 2007.

7. I. Monne, T. M. Joannis, A. Fusaro et al., “Reassortant avian influenza virus (H5N1) in poultry, Nigeria, 2007,” Emerging Infectious Diseases, vol. 14, no. 4, pp. 637–640, 2008.

8. F. O. Fasina, A. C. Meseko, T. M. Joannis et al., “Control versus no control: Options for avian influenza H5N1 in Nigeria,” Zoonoses and Public Health, vol. 54, no. 5, pp. 173–176, 2007.

9. H. Yoon, C. K. Park, H. M. Nam, and S. H. Wee, “Virus spread pattern within infected chicken farms using regression model: the 2003-2004 HPAI epidemic in the Republic of Korea,” Journal of Veterinary Medicine, vol. 52, no. 10, pp. 428–431, 2005.

10. S. B. Muzaffar, J. Y. Takekawa, D. J. Prosser, S. H. Newman, and X. Xiao, “Rice production systems and avian influenza: interactions between mixed-farming systems, poultry and wild birds,” Waterbirds, vol. 33, no. 1, pp. 219–230, 2010.

11. K. Subbarao, A. Klimov, J. Katz et al., “Characterization of an avian influenza A (H5N1) virus isolated from a child with a

fatal respiratory illness," Science, vol. 279, no. 5349, pp. 393–396, 1998.

12. D. F. Adene and A. E. Oguntade, The Structure and Importance of the Commercial and Village based Poultry Industry in Nigeria, Food and Agricultural Organization, Rome, Italy, 2006.

13. L. Saidu, A. M. Wakawa, P. A. Abdu et al., "Impact of avian influenza in some states of Nigeria,"International Journal of Poultry Science, vol. 7, no. 9, pp. 913–916, 2008.

14. P. R. Kumbish, T. M. Joannis, A. R. Jambalang et al., "Clinicopathological features of highly pathogenic Avian Influenza (HP AI-H5N1) outbreaks in commercial chickens in Nigeria," Vom Journal of Veterinary Science, pp. 13–22, 2006.

15. P. de Benedictis, T. M. Joannis, L. H. Lombin et al., "Field and laboratory findings of the first incursion of the Asian H5N1 highly pathogenic avian influenza virus in Africa," Avian Pathology, vol. 36, no. 2, pp. 115–117, 2007.

16. N.V.R.I Records, National Veterinary Research Institute, Vom , Nigeria. Case database, 2006–2008.

17. Food and Agricultural Organization, "FAO Poultry sector country review—Nigeria," Food and Agriculture Organization of the United Nations, Geneva, Switzerland, 2008.

18. [NADIS] National Animal Disease Information and Surveillance Info, Special avian flu edition, National Animal Disease Information and Surveillance (NADIS), 10th edition, 2006.

19. A. Durosinlorun, J. U. Umoh, P. A. Abdu, and I. Ajogi, "Serologic evidence of infection with H5 subtype influenza

virus in apparently healthy local chickens in Kaduna State, Nigeria," Avian Diseases, vol. 54, no. 1, pp. 365–368, 2010.

20. F. O. Fasina, M. M. Sirdar, and S. P. R. Bisschop, "The financial cost implications of the highly pathogenic notifiable avian influenza H5N1 in Nigeria," Onderstepoort Journal of Veterinary Research, vol. 75, no. 1, pp. 39–46, 2008.

CHAPTER 3

Evaluation of Microbiological and Chemical Contaminants in Poultry Farms

Justyna Skóra [1], Katarzyna Matusiak [1,*], Piotr Wojewódzki [2], Adriana Nowak [1], Michael Sulyok[3], Anna Ligocka [4], Małgorzata Okrasa [5], Janusz Hermann [2] and Beata Gutarowska [1]

[1]Faculty of Biotechnology and Food Science, Institute of Fermentation Technology and Microbiology, Lodz University of Technology, 90 924 Łódź, Poland

[2]Department of Environmental Chemistry, University of Technology and Life Science in Bydgoszcz, 85 796 Bydgoszcz, Poland

[3]Department of Agrobiotechnology, Center for Analytical Chemistry, University of Natural Resources and Life Sciences Vienna (BOKU), 3430 Tulln an der Donau, Austria

[4]Department of Microbiology and Food Technology, University of Technology and Life Science in Bydgoszcz, 85 796 Bydgoszcz, Poland

[5]Department of Personal Protective Equipment, Central Institute for Labour Protection, National Research Institute, 00 701 Warsaw, Poland

ABSTRACT

The aim of the study was to evaluate the microbiological and chemical contamination in settled dust at poultry farms. The scope of research included evaluating the contributions of the various granulometric fractions in settled dust samples, assessing microbial contamination using culture methods, concentrations of secondary metabolites in dust and their cytotoxicity against hepatocyte chicken cells by means of MTT (3-(4,5-dimethylthiazolyl-2)-2,5-diphenyltetrazolium bromide) tests. In addition, we also evaluated the concentration of selected volatile odorous compounds (VOCs) using gas chromatographic and spectrophotometric methods and airborne dust concentration in the air with DustTrak™ DRX Aerosol Monitor. Studies were carried out on chicken broilers and laying hens at 13 poultry farms, with numbers of birds ranging from 8000 to 42,000. The airborne total dust concentration at poultry farms averaged 1.44 mg/m^3 with a high percentage of the PM_{10}fraction (particulate matter with a diameter less than 10 μm). Microorganism concentrations in the settled dust were: 3.2×10^9 cfu/g for bacteria and 1.2×10^6 cfu/g for fungi. Potential pathogens (*Enterococcus* spp., *Escherichia coli, Salmonella* spp., *Aspergillus fumigatus*, *Paecilomyces variotii*) were also found. Secondary metabolites included aurofusarin, deoxynivalenol, 15-hydroxyculmorin zearalenone, zearalenone-sulfate, infectopyron, and neochinulin A. However, the dust samples showed weak cytotoxicity towards chicken hepatocyte cells, which ranged between 9.2% and 29.7%. Among volatile odorous compounds ammonia, acrolein, methyloamine, acetic acid, acetoaldehyde and formaldehyde were detected in the air. In conclusion, settled dust can be a carrier of microorganisms, odours and secondary metabolites in poultry farms, which can be harmful to workers' health.

KEYWORDS

poultry farm; microorganisms; air quality; VOCs; mycotoxins; settled dust

1. INTRODUCTION

Animal production can pose serious problems to both workers, animal health, and the surrounding environment. Poultry production may be connected to high concentrations of organic dust, microorganisms in manure, litter, dust and air, and the emission of volatile odorous compounds. Organic dust in poultry farms is a complex mixture of organic and inorganic particles from faecal material/manure, feed, litter, feathers, dander (skin material), mites, bacteria, fungi and fungal spores and endotoxins, depending on the type of bird and stage of the production cycle [1]. Poultry dust may contain bacteria and fungi of plant and animal origin at concentrations ranging between 10^4 and 10^7 cfu/g [2,3]. The presence of mould species: *Acremonium, Alternaria, Aurobasidium Aspergillus*, *Basidiospores, Cladosporium, Chrysosporium, Drechslera, Epicoccum, Eurotium, Fusarium, Geomyces, Mucor, Penicillium, Pithomyces, Rhizomucor, Scopulariopsis* and*Ulocladium* have been reported to be prevalent in poultry dust [4]. Many of them (e.g., *Alternaria, Aspergillus, Fusarium*) are recognised as allergenic strains. Directive 2000/54/EC [5] lists microorganisms in the manure of industrial poultry farms that can potentially be hazardous. Among these, Group 3 organisms include *Bacillus anthracis*, *Chlamydia ornithosis*, *Salmonella choleraesuis var.* Typhi, H5N1 virus; while Group 2 include *Aspergillus fumigatus, Candida albicans, Cryptococcus neoformans*, *Listeria monocytogenes*, *Mycoplasma* spp. *Staphylococcus* spp., *Streptococcus*spp., [4,6]. These and other pathogens may be harmful both to poultry farm workers and birds. Moreover, dust contains biological toxins, including immunotoxicity factors, for example bacterial endotoxin, fungal mycotoxins and glucans, volatile odorous compounds, plant toxins, animal venoms [7,8].

The literature suggests that people working on poultry farms are exposed to higher levels of organic dust compared to those working in cow or swine breeding [9]. Organic dust, which penetrates the respiratory system of workers, may induce toxicity, irritation, allergies, cancer or fibrosis, and result in diseases such as chronic obstructive lung disease, asthma, chronic bronchitis, bronchial hyperreactivity, organic dust toxic syndrome, and irritation of the mucous membranes of the conjunctiva and skin [10,11]. Donham *et al.* [12] reported a decline of lung function in poultry workers. Chronic respiratory symptoms, decreased lung function, rhinitis or eczema were symptoms identified among poultry workers with more than 5 years of occupational exposure [12].

Poultry farms are the biggest emitters of volatile odorous compounds. On poultry farms these are the result of fermenting and rotting litter, and also due to the decomposition of manure and debris. Part of the gas may result from breathing, digestion and release from the skin of the birds. Research is still ongoing to identify all odorous gases in livestock with current estimates identifying approximately 130 compounds. These include hydrogen sulphide, ammonia, thiols, sulphides and aliphatic amines, heterocyclic organic compounds containing sulphur and nitrogen, aliphatic alcohols and phenols, ketones, aldehydes, aliphatic acids, and esters [13]. According to the Reference Document on Best Available Techniques for Intensive Rearing of Poultry and Pigs (ILF BREF), and The European Pollutant Release and Transfer Register (E-PRTR), the main pollutants emitted from poultry houses are ammonia (NH_3), methane (CH_4), nitrogen monoxide (N_2O), and inspirable and respirable dust [14,15]. Moreover, odorous emissions from poultry houses are highly variable. It depends on numerous factors including the age of the birds, the season, dietary composition, temperature, humidity, litter type and quantity, manure handling operations, bird density, and the type and rate of ventilation [16,17,18,19,20,21].

Nevertheless, legislations concerning environmental protection (Directive 2008/50/EC, 2008) and animal welfare (Council Directive 2007/43/EC, 2007) set limits on the concentration of only certain substances: carbon dioxide (3000 ppm) and ammonia (20 ppm) [22,23]. In addition, there are no established regulations on the permissible levels of settled dust in livestock, types of microorganisms and their secondary metabolites. Therefore, it is necessary to undertake additional research on the microbiological and chemical characteristics of dust present in poultry farms, which may be a health threat to workers and animals. Currently, there is limited knowledge on the cytotoxicity of settled dust in poultry farms towards chicken hepatocytes cells.

The aim of the study was to evaluate the microbiological and chemical contamination in settled dust in poultry farms. The scope of research included evaluating the contributions of various granulometric fractions in settled dust samples, assessing microbial contamination, determining the concentrations of secondary metabolites and their cytotoxicity against hepatocyte chicken cells. In addition, we also evaluated concentrations of selected odorous volatile compounds in the air of the poultry farms.

2. MATERIALS AND METHODS

2.1. Poultry Farms and Dust Samples

Analysis was performed between weeks 3 and 57 of the breeding cycle (farms I-X—broilers kept on deep litter systems; farms XI-XIII—laying hens kept in caged systems) at 13 poultry farms located in Kuyavia-Pomerania and Lodz districts (Poland) containing from 8000 to 42,000 birds.Table 1 summarizes the characteristics of the livestock buildings.

Table 1. Livestock building descriptions.

Poultry Farm	Description	Birds
I [B]	Area: 1831 m^2, feeders (4 lines), 5 drip-type drinker lines, 12 roof-mounted fans, 4 wall-mounted fans	37,500
II [B]	Area: 1260 m^2, feeders (2 lines), 3 drip-type drinker lines, 9 roof-mounted fans, 4 wall-mounted fans	25,000
III [B]	Area: 1750 m^2, feeders (3 lines), 4 drip-type drinker lines, 18 roof-mounted fans, 4 wall-mounted fans	33,000
IV [B]	Area: 2100 m^2, feeders (4 lines), 5 drip-type drinker lines, 18 roof-mounted fans, 13 wall-mounted fans	42,000
V [B]	Area: 1566m^2, feeders lines (5 lines), 6 drip-type drinker lines, 11 roof-mounted fans	30,000
VI [B]	Area: 1383m^2, feeders (3 lines), 4 drip-type drinker lines, 16 roof-mounted fans, 2 wall-mounted fans	27,400
VII [B]	Area. 1085m^2, feeders (3 lines), 4 drip-type drinker line, 6 roof-mounted fans, 3 wall-mounted fans	18,000
VIII [B]	Area: 1750 m^2, feeders (3 lines), 4 drip-type drinker lines, 18 roof-mounted fans, 4 wall-mounted fans	33,600
IX [B]	Area: 1831 m^2, feeders (4 lines), 5 drip-type drinker lines, 12 roof-mounted fans, 4 wall-mounted fans	37,500
X [B]	Area: 1074 m^2, feeders (3 lines), 4 drip-type drinker lines, 11 wall-mounted fans	24,000
XI [H]	Area: 1430 m^2, feeders (3 lines), 4 drip-type drinker lines, 11 wall-mounted fans	8000
XII [H]	Area: 1080 m^2, feeders lines (2 lines), 4 drip-type drinker lines, 10 roof-mounted fans, 2 wall-mounted fans	8000
XIII [H]	Area: 1200 m^2, feeders (lines 6), 7 drip-type drinker lines, 6 roof-mounted fans, 4 wall-mounted fans	27,800

B—broilers on deep litter systems; H—laying hens in caged systems.

Settled dust is the part of airborne particulate matter that fell down onto sampling surface during the chicken broiler production cycle. Three metal plates were set at a height of about 1.6 m, evenly in both ends and in the middle of every analyzed poultry house. The sampling plates were placed in the poultry buildings the day before introducing chickens and on the last day of production cycle, after removing chickens from the building, the sampling plates containing settled dust were collected. Settled dust samples were collected after the production cycle on three sampling surfaces (metal plates; surface dimensions listed in Table 2) in each chicken broiler house. Samples were carefully swept with disposable brushes into polypropylene string bags, then mixed together and used for further analyses. The average mass (weight (g)) of dust collected is also presented in the Table 2. Microbial contamination, secondary metabolites and cytotoxicity were determined in the dust samples (farms I-X), and airborne dust concentration in farm XIII. Analyses of selected volatile odorous compounds in the air were undertaken for poultry farm II at three stages of the production cycle: stage I (without chickens); stage II (20th day), stage III (35th day), and stages XI and XII (57th week, at the end of the production cycle). The research was carried out between January and July 2015.

2.2. Airborne Dust

Airborne dust concentration was measured using a portable laser photometer DustTrak™ DRX Aerosol Monitor 8533 (TSI, Shoreview, MN, USA), which allows simultaneous measurements of size-segregated mass fractions corresponding to PM_1 (Particulate Matter), $PM_{2.5}$, PM_4 (respirable), PM_{10} and total PM size fractions. Measurements were carried out five times for 15 min with a sampling interval of 1 s (the number of samples for each measurement was n = 900). Dust concentration data were then used to calculate 8 h equivalent time-weighted averages, but as the measurements only lasted for short periods, the calculated values should be treated as estimates [24].

2.3. Size Fractions of Settled Dust

Samples of settled dust from three sampling points in each poultry farm were taken for granulometric fraction analysis (the total number of samples was 60). A laser diffraction dust particle size analyser, Mastersizer 2000 ver. 5.60, equipped with a Hydro 2000MU wet sample dispersion unit (Malvern Instruments, Ltd.; Malvern, UK), was used for various granulometric fraction analyses of the settled dust samples. The dust fraction $PM_{2.5}$ (respirable dust with a diameterless than 2.5 µm), PM_{10} (particulate matter with a diameter less than 10 µm) and PM > 10 (particulate matter with a diameter more than 10 µm) were each analysed in triplicates and presented as the percentage of settled dust.

2.4. Microbial Contamination

Samples of settled dust from each poultry farm were analysed microbiologically. For this purpose, 4–32 g samples of settled dust were collected in sterile bins, mixed, and a 1 g sample was suspended in 99 mL saline solution (0.85% NaCl). The samples were diluted from 10^{-1} to 10^{-8} in duplicates and plated on MEA (Merck Millipore, Darmstadt, Germany) medium with (0.1%) chloramphenicol (fungi); Nutrient agar (Merck) with (0.2%) nystatin (bacteria); Actinomycete Isolation Agar (Hi Media Laboratories, Maharashtra, India) with (0.2%) nystatin (actinomycetes); Chapman agar (mannitol-positive *Staphylococcus* spp.), King B medium (Hi Media Laboratories) (*Pseudomonas fluorescens*), ENDO medium (Merck) (*Escherichia coli*); Kanamycin Aesculin Azide medium (Merck) (*Enterococcus* spp.); BPLA medium, following Kaufmann (Merck) (*Salmonella*spp.) Samples were incubated at 37 ± 2 °C for 24–48 h (total number of bacteria, mannitol-positive*Staphylococcus* spp., *Escherichia coli, Enterococcus* spp., *Salmonella* spp.), at 27 ± 2 °C for 5–7 days (fungi, actinomycetes), or at 30 ± 2 °C for 48 h (*Pseudomonas fluorescens*). After incubation, the colonies were counted, and the results were expressed in cfu/g of dust. Data was analysed from three independent experiments. The final result was calculated as the arithmetic mean and standard deviation (SD) of all repetitions.

The pure cultures of yeast and moulds were characterized macroscopically, and then, identified. For yeasts, diagnostics was performed using the API C AUX test (BioMérieux, Marcy-l'Étoile, France). Isolated moulds were identified by macroscopic and microscopic observations of culture on CYA (czapek yeast extract agar) and YES (yeast extract with supplements) media, using taxonomic keys [25,26,27,28]. Percentage of isolated mould colonies and the frequency of isolation (the percentage of samples in which isolated the species) for the identified species were determined.

2.5. Secondary Metabolites

Secondary metabolite analyses were performed for 10 settled dust samples taken from each tested poultry farm. For this purpose 4–32 g of dust were collected in sterile bins and mixed. Next, 0.5 g of dust samples were suspended in 5 mL of the extraction solvent (acetonitrile/water/acetic acid 79:20:1, v:v:v). Samples were extracted for 90 min and diluted with the same volume of solvent prior to injection [29] Secondary metabolite concentrations were analysed quantified using LC-MS/MS, as described by Sulyok *et al.* [30] with further modification. Parameters for liquid chromatography and mass spectrometry are described elsewhere [31]. Briefly, LC-MS/MS screening of target microbial metabolites was performed with a QTrap 5500 LC-MS/MS System (Applied Biosystems, Foster City, CA, USA) equipped with a TurboIonSpray electrospray ionization (ESI) source and a 1290 Series HPLC System (Agilent, Waldbronn, Germany). Chromatographic separation was performed at 25 °C on a Gemini® C18-column, 150 × 4.6 mm i.d., 5 mm particle size, equipped with a C18 4 × 3 mm i.d. security guard cartridge (Phenomenex, Torrance, CA, USA). ESI-MS/MS was performed in the time-scheduled multiple reaction monitoring (MRM) mode, both in positive and negative polarities, in two separate chromatographic runs per sample, by scanning two fragmentation reactions per analyte. The MRM detection window of each analyte was set to its expected retention times of ±27 and ±48 s in the positive and negative modes, respectively. The positive analyte was confirmed by the acquisition of two MRMs per analyte, with the exception of moniliformin, which exhibited only one fragment ion. This yielded 4.0 identification points according to the European Union Commission decision 2002/657 [32]. The LC retention time and the intensity ratio of the two MRM transitions agreed with the related values of an authentic standard within 0.1 min and 30% rel. Analysis was performed in three replicates per sample.

2.6. Cytotoxicity

Cytotoxicity analysis was performed for 10 settled dust samples taken from each tested poultry farm. For this purpose, 4–32 g of dust were collected in sterile bins and mixed. Next, water-soluble fraction extracts were prepared: 0.1 g of dust samples were suspended in 20 mL of PBS—Phosphate Buffered Saline (Sigma-Aldrich, St. Louis, MO, USA) (dust concentration in extracts was

0.5%), pH7.2. Samples were extracted for 30 min, next 3 mL of each extract was filtered by sterile syringe filters (0.22 µm) (MerckMillipore, Darmstadt, Germany). Cytotoxicity of the prepared extracts was determined using the MTT assay with chicken liver hepatocellular carcinoma cell line, Leghorn male hepatoma (LMH) (CLS, Eppelheim, Germany, lot no. 601411-714SF), after 24 passages.

Once the cells became adherent they were cultured in collagen coated Roux flasks (BioCoat, Becton, Dickinson and Co., Franklin Lakes, NJ, USA) as a monolayer in Waymouyh's Medium (Gibco, Thermo Fisher Scientific, Waltham, MA, USA) with the addition of 7.5% sodium bicarbonate (Gibco), 10% FBS heat-inactivated foetal bovine serum (Gibco), 25 mM HEPES (Sigma-Aldrich), 100 IU/mL penicillin (Sigma-Aldrich), and 100 µg/mL streptomycin (Sigma-Aldrich). The cells were incubated in a CO_2 incubator at 37 °C in 5% CO_2 for 7 days to reach 80% confluency. The medium was changed every 3–4 days. After reaching confluence, the cells were sub-cultured. They were detached with TrypLE™ Express (Gibco) for 5 min at 37 °C, suspended in sterile PBS and aspirated off the plastic flask. As the enzyme is of plant origin the reaction does not need to be terminated with FBS. Following detachment, the cell suspension was transferred to a 15 mL Falcon tube, centrifuged (182× *g*, 5 min), decanted and resuspended in fresh medium. After determination of cell count and viability by trypan blue exclusion (minimus 90%), the cells were ready to use.

In the experiment, 1×10^4 LMH cells were placed in each well of a collagen coated 96-well plate (BioCoat) and 100 µL of the complete culture medium was added to each well. The cells were incubated overnight at 37 °C in 5% CO_2 to allow them to attach. The following day, settled dust extracts (0.5%) were diluted in Waymouth's Medium with no FBS (1:4). The final concentrations of dust extract was 0.125%. The medium was aspirated from the 96-well culture and 200 µL of each tested sample was added to well in eight repeats. The control samples consisted of cells without dust extracts. The cells were incubated in a CO_2 incubator at 37 °C with 5% CO_2 for 48 h.

Table 2. Settled dust sample characteristics.

Poultry Farm	Collection Surface (m^2)	Weight (g) *	Humidity (%)	Fall (g/cm^2)
I	1.14 ± 0.00	21.32 ± 6.54	23.35 ± 8.07	0.0019 ± 0.0006
II	1.01 ± 0.00	9.81 ± 1.25	19.13 ± 3.41	0.0010 ± 0.0001
III	0.86 ± 0.00	32.63 ± 15.79	8.62 ± 0.43	0.0038 ± 0.0018
IV	0.86 ± 0.00	25.10 ± 16.34	8.85 ± 0.26	0.0029 ± 0.0019
V	0.32 ± 0.00	8.18 ± 3.57	6.52 ± 0.36	0.0031 ± 0.0023
VI	0.37 ± 0.00	13.29 ± 0.29	6.28 ± 0.47	0.0036 ± 0.0010
VII	0.33 ± 0.00	13.13 ± 0.21	8.58 ± 0.41	0.0039 ± 0.0010
VIII	0.20 ± 0.00	8.34 ± 5.08	6.32 ± 0.49	0.0042 ± 0.0026
IX	0.09 ± 0.00	5.57 ± 0.44	10.81 ± 0.31	0.0061 ± 0.0050
X	0.09 ± 0.00	4.29 ± 0.21	10.42 ± 0.54	0.0047 ± 0.0020

* average mass of dust collected on three sampling plates from each chicken broiler house.

After incubation, the medium with tested dust extracts was gently aspirated from each well and 100 μL of MTT (0.5 mg/mL in PBS; Sigma-Aldrich, St. Louis, MO, USA,) was added and incubated at 37 °C in 5% CO_2 for 3 h. Following incubation, MTT was carefully removed and formazan precipitates were solubilised with 50 μL of DMSO (Sigma-Aldrich, St. Louis, MO, USA). Absorbance was measured at 550 nm with a reference filter of 620 nm, using a TriStar^2 LB 942 microplate reader (Berthold Technologies, Bad Wildbad, Germany). The absorbance of the control sample (untreated cells with dust extract) was taken to represent 100% cell viability. Cell viability (%) was calculated as follows: (sample OD (optical density)/control OD) × 100%; and cytotoxicity (%) as: 100-cell viability (%). Results were presented as means ± SD (standard deviation).

2.7. Volatile Odourous Compounds

Analysis of air quality included the measurement of the following volatile odorous compound concentrations: ammonia, methylamine, dimethylamine, trimethylamine, hydrogen sulfide, methyl mercaptan, ethyl mercaptan, propyl mercaptan, butyl carbon monoxide, carbon dioxide, oxygen, total organic carbon, acrolein, acetaldehyde, formaldehyde formic acid and acetic acid.

Air samples (0.02–0.06 m^3) were collected in tedlar bags using an aspirator (EAS 1203; Emio, Wrocław, Poland). Analyses of the collected compounds were performed by GC/MS (Perkin Elmer Arnel 1115, Waltham, MA, USA and Bruker Daltonics Inc. 436-GC, Manning Park Billerica, MA USA) equipped with an RTX-1 column, PFPD, TCD and FID detectors, and using nitrogen and helium as carrier gases. The study was performed according to Polish standards (PN) (Table 3). Results were calculated for the conditions of 1013 hPa and 293 K.

Table 3. Detection standards.

No.	VOCs	Norm
1	acetaldehyde	ZBES/PB/17 eighth edition of 07.02.2012
2	acetic acid	ZBES/PB/16 seventh edition of 07.02.2012
3	acrolein	ZBES/PB/17 eighth edition of 07.02.2012
4	ammonia	PN-EN ISO 11732:2007
5	butyl mercaptan	PB GE 18 third edition of 16.02.2009
6	carbon dioxide	PB GE 22 (first edition of 03.02.2010)
7	carbon monoxide	
8	diethylamine	ZBES/PB/32 third edition of 02.01.2013
9	dimethylamine	
10	ethyl mercaptan	PB GE 18 third edition of 16.02.2009
11	ethylamine	ZBES/PB/32 third edition of 02.01.2013
12	formaldehyde	ZBES/PB/17 eighth edition of 07.02.2012
13	formic acid	PN-88/Z-04196.02
14	hydrogen sulfide	PB GE 18 third edition of 16.02.2009
15	methyl mercaptan	PB GE 18 third edition of 16.02.2009
16	methylamine	ZBES/PB/32 third edition of 02.01.2013
17	oxygen	PB GE 22 (first edition of 03.02.2010)
18	propyl mercaptan	PB GE 18 third edition of 16.02.2009
19	total organic carbon	PN-EN 13526:2005; PN-EN 12619:2002
20	triethylamine	ZBES/PB/32 third edition of 02.01.2013
21	trimethylamine	

Air samples were also collected in the scrubber at a flow rate of 0.6 m^3/h. Ammonia and formic acid were determined using a spectrophotometer (CFA SAN spectrophotometer, SKALAR, Breda, Netherlands), according to the norms (Table 3). Total organic carbon was determined according to the norms (Table 3) using a portable, heated FID total hydrocarbon analyzer, OVF-3000 (J.U.M.®Engineering GmbH, Karlsfeld, Germany).

2.8. Statistical Analysis

Statistical analyses was performed for number of selected microorganisms in settled dust as well as for share of granulometric fractions of dust and cytotoxicity properties of dust from different poultry farms. Statistical analyses were conducted using STATISTICA 10 software (Statsoft, Round Rock, TX, USA). Descriptive statistics for all variables of interest were calculated. The results were evaluated using one-way analysis of variance (ANOVA) at the significance level 0.05. When statistical difference was detected ($p < 0.05$), means were compared using Tukey's post hoc procedure at the significance level 0.05.

3. RESULTS AND DISCUSSION

In our study, the highest concentration of airborne dust was found in the PM_{10} dust fraction (average: 0.875 mg/m^3, maximum: 2.128 mg/m^3) in poultry farms. Dust particles with lower diameters PM_1, $PM_{2.5}$, PM_4 were found at levels 0.480–0.541 mg/m^3 (Table 4).

Table 4. Airborne dust concentration in poultry farm.

Dust Fraction	Concentration (mg/m^3)
PM_1	X: 0.480 ± 0.113 Min: 0.201 ± 0.046 Max: 1.080 ± 0.165
$PM_{2.5}$	X: 0.493 ± 0.114 Min: 0.206 ± 0.047 Max: 1.101 ± 0.165
PM_4	X: 0.541 ± 0.116 Min: 0.227 ± 0.054 Max: 1.189 ± 0.159
PM_{10}	X: 0.875 ± 0.121 Min: 0.352 ± 0.094 Max: 2.128 ± 0.177
PM_{total}	X:1.440 ± 0.132 Min: 0.502 ± 0.143 Max: 3.716 ± 0.431

PM_1; $PM_{2.5}$; PM_4; PM_{10}—respectively: granulometric fraction with diameter less than 1; 2.5; 4; 10 μm; X—mean; Min, Max—minimum and maximum value.

Ellen *et al.* [33] and Viegas *et al.* [34] carried out studies to determine dust concentrations in poultry farms. The authors found similar results, where PM_{10} predominated, albeit at higher concentrations (1.4–15.2 mg/m^3). Concentration of airborne dust may depend on the type of bird (lying hens, broilers), the stage of the production cycle, the ventilation system in the poultry farm and other factors [33]. Granulometric fractions with particles of diameters <2.5 µm made up 1.0%–1.9% of settled dust (Table 5).

Dust particles smaller than 10 µm made up 6.8 to 10.4% of the settled dust. Significantly higher ($p < 0.05$) percentages of both fractions (2.7% and 13.7%, respectively) were observed in farm II for $PM_{2.5}$ and PM_{10}. The percentage of particles with diameter more than 10 µm at farm II amounted to 86.3% and was significantly lower ($p < 0.05$) compared to others farms. However, most of the settled dust particles analysed were above 10 µm (89.6%–93.2%) (Table 5). This fact indicates the dominance of the same granulometric size fractions of dust in both the settled dust and the air. The dominance of this fraction is probably associated with a particular poultry farm's specificity and dust composition—feathers, skin of birds, litter and feed particles falling from the air to the surface. Dust particles >10 µm are not capable of deep penetration into the human respiratory system, and are mostly deposited in the mouth and nose [35]. Exposure to PM_{10} may also produce disease by impacting the upper and larger airways below the vocal cord [34].

Table 5. Percentage share of granulometric fractions of settled dust.

Poultry Farm	Fractions of Settled Dust (%)		
	$PM_{2.5}$	PM_{10}	PM > 10
I	1.88 ± 0.26 ^{B}	10.10 ± 1.04 ^{B}	89.91 ± 1.04 ^{D}
II	2.69 ± 0.09 ^{A}	13.71 ± 0.42 ^{A}	86.29 ± 0.42 ^{E}
III	1.92 ± 0.32 ^{B}	10.42 ± 1.07 ^{B}	89.58 ± 1.07 ^{D}
IV	1.81 ± 0.38 ^{B,C}	10.17 ± 1.56 ^{B}	89.83 ± 1.56 ^{D}
V	1.43 ± 0.06 ^{B,C,D}	9.84 ± 0.08 ^{B,C}	90.17 ± 0.07 ^{C,D}
VI	1.42 ± 0.01 ^{B,C,D}	9.55 ± 0.10 ^{B,C,D}	90.45 ± 0.10 ^{B,C,D}
VII	1.14 ± 0.06 ^{D}	7.37 ± 0.30 ^{D,E}	92.63 ± 0.30 ^{A,B}
VIII	1.27 ± 0.14 ^{C,D}	7.69 ± 0.69 ^{C,D,E}	92.31 ± 0.68 ^{A,B,C}
IX	0.97 ± 0.08 ^{D}	6.85 ± 0.35 ^{E}	93.15 ± 0.35 ^{A}
X	1.06 ± 0.16 ^{D}	7.26 ± 0.63 ^{E}	92.74 ± 0.63 ^{A}
Mean	1.56 ± 0.52	9.30 ± 2.08	90.70 ± 2.08

$PM_{2.5}$—granulometric fraction with diameter less than 2.5 μm; PM_{10}—granulometric fraction with a diameter less than 10 μm; PM—granulometric fraction with a diameter more than 10 μm, $^{A, B, C, D, E}$—means with the same capital letter in the same column are not significantly different (Tukey's test; $p < 0.05$).

It is worth emphasizing that there are no regulations/limits for the percentage of each fraction in the settled dust in workplaces. Existing recommendations directly relate to concentrations of particulate matter in the air. According to the Approved Code of Practice (ACOP) the Control of Substances Hazardous to Health Regulations 2002 (COSHH) [36] any kind of dust can become hazardous to health when it is present at concentrations 10 mg/m^3 or higher in the air (weighted arithmetic mean during eight-hour period) for inhalable dust, or 4 mg/m^3 (weighted arithmetic mean during eight-hour period) for respirable dust. In our study, airborne dust did not exceed the recommended limits. It is worth emphasizing that the limits take into account single agent exposure. However in the tested working environments there exists co-exposure to several physical (including dust), chemical (odours) and biologic agents (pathogens).

The different groups of microorganisms in the settled dust of the poultry farms are shown inTable 6. Bacteria (average: 3.2×10^9 cfu/g), and actinomycetes (5.5×10^6 cfu/g) dominated all isolated microorganisms, while fungi occurred less frequently (1.2×10^6 cfu/g). The total number of bacteria in settled dust samples varied depending on the poultry farms tested, and ranged from 1.5×10^7 to 2.9×10^9 cfu/g; significantly different ($p < 0.05$) bacteria number was detected in IV poultry farm 1.7×10^{10} cfu/g. The concentration of actinomycetes in tested dust samples was at a similar level, with the exception of farms I and IV which showed a statistically significant difference in their number ($p < 0.05$). In all tested dust samples, mannitol positive *Staphylococci* spp. (1.7×10^9cfu/g) and *Enterococcus* spp. (5.8×10^7 cfu/g) were identified. *E. coli* was present in 90% of the analysed dust samples (with concentration 1.6×10^5 cfu/g). *Pseudomonas fluorescens* was isolated from 4/10 dust samples (8.7×10^5 cfu/g). A particularly high number of these bacteria was detected at poultry farm IV ($p < 0.05$). Only in 3/10 dust samples were *Salmonella* spp. identified

(9.5×10^4cfu/g). A statistically significant number of bacteria belonging to *Salmonella* genera was detected on farm VIII ($p < 0.05$) (Table 6).

Baker *et al.* [37] and Witkowska *et al.* [38] found that the number of aerobic bacteria in litter were at a level of 10^7–10^9 cfu/g; (from 9.5×10^7 cfu/g, which increased to 2.3×10^9 cfu/g after 5 weeks of broiler settlement). We found similar bacterial genera in settled dust in our study as described in other studies on poultry farms in literature: *Bacillus*, *Clostridia*, *Corynebacterium*, *Enterobacter*, *Flavobacterium*, *Pseudomonas*, *Staphylococcus*, *E. coli*. [39].

According to Directive 2000/54/EC [5] *Enterococcus*, *Escherichia coli* and *Salmonella* that were isolated in our study, can be harmful to workers. Dutkiewicz *et al.* [6] also indicated that *P. fluorescens* is potentially dangerous to people with an impaired immune system. Moreover, they also found the following potentially pathogenic bacteria in study: *Chlamydia ornithosis*, *Bacillus anthracis*, *Salmonella* Choleraesuis var. Typhi, *Listeria monocytogenes*, *Mycoplasma* spp.,*Staphylococcus aureus*, *Streptococcus* spp. which can occur in manure, litter, settled dust. We also isolated 13 strains of fungi in settled dust, which included three strains of yeast and 11 mould strains (Table 7).

The following strains were isolated at the highest frequencies: *Aspergillus penicillioides* (72.5%),*Eurotium chevelierii* (32.5%), *Mucor fragilis* (27.5%) *Absidia glauca* (27.5%). These species also had the highest percentage of isolations (quantitatively determined) in dust samples, ranging between 5.3% and 29.1% among the isolated colonies (Table 7). Mould species isolated from settled dust samples were saprophytic organisms and are specific to environments such as soil or plant material [28]. According to the literature, the most common fungal genera reported in settled dust include *Cladosporium*, *Penicillium*, *Aspergillus*, although *Alternaria*, *Fusarium* and *Geotrichum* can also be present [39].

A. fumigatus, one of the moulds identified in settled dust samples in two poultry farms, belongs to the second health risk group according to Directive 2000/54/EC [5], which concerns protecting workers from risks related to biological agents in work places. This species is known for its infection ability, allergenic and toxinogenic properties [27]. Also, *Paecilomyces variotii*, a species with a relatively pronounced ability to survive in vertebrate tissue, is classified to BSL2 (biosafety level 2) by the European Confederation of Medical Mycology (ECMM) [40]. In severely immunocompromised people, it may cause serious opportunistic mycoses [40]. The literature also identifies other potentially patogenic species from poultry farm environments: *Candida albicans*,*Cryptococcus neoformans*, which we did not find in our study. However, we identified another species from the genera *Candida* (*C. pelliculosa*) and *Cryptococcus* (*C. uniguttulatus*).

In the settled dust samples, 27 chemicals classified as secondary mould metabolites were identified (Table 8). Metabolites identified in high concentrations were: aurofusarin, deoxynivalenol, 15-hydroxyculmorin zearalenone, zearalenone-sulfate (VIII), infectopyron (VIII–X), and neochinulin A (IX, X). Asperglaucide, brevianamid F, enniatin B, enniatin B1 were detected in all samples. Other compounds varied by the poultry farm where the dust sample was isolated.

The highest numbers of secondary metabolites were detected in the dust sample isolated from poultry farm VIII. It included typical *Fusarium* spp. mycotoxins: 15-Hydroxyculmorin apicidin, aurofusarin, α-zearalenol, β-zearalenol, deoxynivalenol, *epi*-equisetin, equisetin, zearalenone and zearalenone-sulfate. Among the moulds isolated, the genus *Fusarium* was not detected. This may be due to previously contaminated plant material e.g., feed for poultry.

Table 6. Number of microorganisms in settled dust in poultry farms.

Poultry Farm	Microorganisms Number (cfu/g, Mean ± SD)							
	Total Number of Bacteria	Actinomycetes	Mannitol Positive *Staphylococci* sp.	*Enterococcus* spp.	*E. coli*	*Salmonella* spp.	*Pseudomonas fluorescens*	Fungi
I	$2.8 \times 10^9 \pm 1.2 \times 10^9$ B,C	$1.4 \times 10^7 \pm 7.8 \times 10^6$ B	$2.6 \times 10^9 \pm 1.4 \times 10^9$ A,B,C	$1.2 \times 10^8 \pm 4.2 \times 10^7$ A	nd	nd	$5.3 \times 10^4 \pm 8.0 \times 10^3$ B	$1.8 \times 10^5 \pm 4.6 \times 10^4$ C
II	$2.9 \times 10^9 \pm 1.5 \times 10^9$ B,C	$3.4 \times 10^5 \pm 6.1 \times 10^4$ C	$4.4 \times 10^9 \pm 1.9 \times 10^9$ A,B	$1.6 \times 10^7 \pm 5.0 \times 10^6$ B	$9.7 \times 10^3 \pm 5.1 \times 10^3$ B	nd	$1.2 \times 10^6 \pm 6.0 \times 10^5$ B	$1.1 \times 10^5 \pm 7.1 \times 10^4$ C
III	$5.5 \times 10^9 \pm 6.0 \times 10^9$ B	$6.0 \times 10^5 \pm 1.1 \times 10^5$ C	$1.7 \times 10^9 \pm 7.5 \times 10^8$ B,C	$1.8 \times 10^8 \pm 9.5 \times 10^7$ A	$5.8 \times 10^5 \pm 1.2 \times 10^5$ A	nd	nd	$1.5 \times 10^6 \pm 5.4 \times 10^5$ B,C
IV	$1.7 \times 10^{10} \pm 4.5 \times 10^9$ A	$7.7 \times 10^4 \pm 4.5 \times 10^4$ C	$5.1 \times 10^9 \pm 2.0 \times 10^9$ A	$1.4 \times 10^8 \pm 1.2 \times 10^7$ A	$2.1 \times 10^4 \pm 9.6 \times 10^3$ B	nd	$7.3 \times 10^6 \pm 4.0 \times 10^6$ A	$3.6 \times 10^5 \pm 1.1 \times 10^5$ C
V	$7.1 \times 10^8 \pm 1.2 \times 10^8$ C	$3.3 \times 10^7 \pm 7.5 \times 10^6$ A	$2.4 \times 10^8 \pm 7.0 \times 10^7$ C	$1.3 \times 10^7 \pm 4.2 \times 10^6$ B	$2.0 \times 10^5 \pm 8.9 \times 10^4$ B	nd	nd	$1.6 \times 10^6 \pm 2.6 \times 10^5$ A,B,C
VI	$4.4 \times 10^8 \pm 7.0 \times 10^7$ C	$2.4 \times 10^6 \pm 7.0 \times 10^5$ C	$6.2 \times 10^7 \pm 1.2 \times 10^7$ C	$9.7 \times 10^6 \pm 4.0 \times 10^6$ B	$2.5 \times 10^5 \pm 8.0 \times 10^4$ B	nd	nd	$5.6 \times 10^5 \pm 2.7 \times 10^5$ C
VII	$4.1 \times 10^8 \pm 6.6 \times 10^7$ C	$1.1 \times 10^6 \pm 4.5 \times 10^5$ C	$1.8 \times 10^8 \pm 3.1 \times 10^7$ C	$1.2 \times 10^7 \pm 4.5 \times 10^6$ B	$2.0 \times 10^4 \pm 1.2 \times 10^4$ B	nd	nd	$1.0 \times 10^6 \pm 3.4 \times 10^5$ B,C
VIII	$1.5 \times 10^7 \pm 6.8 \times 10^6$ C	$1.5 \times 10^6 \pm 4.6 \times 10^5$ C	$1.6 \times 10^6 \pm 5.7 \times 10^5$ C	$1.5 \times 10^6 \pm 4.5 \times 10^5$ B	$7.0 \times 10^3 \pm 4.0 \times 10^3$ B	$6.3 \times 10^5 \pm 3.1 \times 10^5$ A	$8.0 \times 10^4 \pm 4.0 \times 10^4$ B	$4.6 \times 10^5 \pm 1.8 \times 10^5$ C
IX	$2.0 \times 10^9 \pm 4.2 \times 10^8$ B,C	$1.2 \times 10^6 \pm 2.1 \times 10^5$ C	$1.6 \times 10^9 \pm 5.7 \times 10^8$ B,C	$8.0 \times 10^7 \pm 3.0 \times 10^7$ A,B	$5.5 \times 10^5 \pm 2.3 \times 10^5$ A	$1.1 \times 10^5 \pm 6.0 \times 10^4$ B	nd	$2.6 \times 10^6 \pm 8.5 \times 10^5$ A,B
X	$8.5 \times 10^8 \pm 6.5 \times 10^7$ C	$1.0 \times 10^5 \pm 2.0 \times 10^4$ C	$8.2 \times 10^8 \pm 7.8 \times 10^7$ C	$8.0 \times 10^6 \pm 3.6 \times 10^6$ B	$5.3 \times 10^3 \pm 2.5 \times 10^3$ B	$2.1 \times 10^5 \pm 1.0 \times 10^5$ B	nd	$3.2 \times 10^6 \pm 1.9 \times 10^6$ A
Mean ± SD	$3.2 \times 10^9 \pm 5.0 \times 10^9$	$5.5 \times 10^6 \pm 1.1 \times 10^7$	$1.7 \times 10^9 \pm 2.0 \times 10^9$	$5.8 \times 10^7 \pm 7.0 \times 10^7$	$1.6 \times 10^5 \pm 2.3 \times 10^5$	$9.5 \times 10^4 \pm 2.1 \times 10^5$	$8.7 \times 10^5 \pm 2.5 \times 10^6$	$1.2 \times 10^6 \pm 1.1 \times 10^6$

SD—standard deviation; nd—not detected in 1 g of settled dust, $^{A, B, C}$—means with the same capital letter in the same column are not significantly different (Tukey's test; $p < 0.05$).

Table 7. Fungal species isolated from settled dust samples.

No	Species	Frequency of Isolation (%)	Percentage of Isolation (%)	Isolation Place (Poultry Farm)
1	*Absidia glauca*	27.5%	7.2%	I–III; V; VII; X
2	*Alternaria alternata*	10.0%	0.8%	III; VI
3	*Aspergillus fumigatus*	15.0%	5.3%	V; VI
4	*Aspergillus penicillioides*	72.5%	32.5%	I–X
5	*Candida pelliculosa* [A]	15.0%	2.3%	III; VII; VIII
6	*Cephaliophora tropica*	5.0%	0.0%	II
7	*Chaetomium globosum*	15.0%	0.7%	II; VI; X
8	*Cryptococcus uniguttulatus* [B]	25.0%	8.5%	I–VI
9	*Eurotium chevelierii*	32.5%	15.2%	IV–V; IX; X
10	*Mucor fragilis*	27.5%	20.2%	II; IV; V; VII; VIII; X
11	*Mucor piriniformis*	17.5%	2.1%	I; II; V; VI; VIII
12	*Paeciliomyces variotii*	60.0%	4.0%	I–X
13	*Rhodotorula mucilaginosa* [C]	27.5%	1.2%	I–V; VIII

[A, B, C]—compatibility of the identification of yeasts obtained by APIweb system respectively: 99,8%; 99,9%; 99,9%.

Dust from poultry farms contains low levels of mycotoxins (e.g., trichothecene B., deoxynivalenol, zearalenone) [41]. Literature data confirms that secondary metabolites of the*Fusarium* genus are a common feed contamination. The most common mycotoxins found in grains for poultry feed are deoxynivalenol (DON) and its derivatives. [42]. In the studies presented, DON was identified at a poultry farm (VIII) at a concentration of 151.2 mg/kg dust. High exposure to DON can lead to nutrition disorders, metabolism and immunity [43]. In all poultry farms, Zearalenone (ZEN) was detected at 0.7–72.2 mg/kg dust. ZEN derivatives were also detected on farms VII and VIII–X: zearalenone-sulfate (42.7–204. 5 μg/kg) and α-zearalenol (36.9 μg/kg). It is worth noting that ZEN has relatively low toxicity, with an LD_{50} value of 2 ± 10 g/kg body weight determined for mice [44]. Its role as a mammalian endocrine disrupter and its genotoxicity potential was previously reported [45].

In this study, characteristic metabolites of the following genera were detected in settled dust: *Alternaria* (alternariol, altersetin, infectopyrone), *Aspergillus* (3-nitropiopropionic acid), *Penicillium* (mycophenolic acid). Also nonspecific metabolites—whose occurrence cannot be linked to specific moulds (asperglaucide, brevianamid F, citreorosein, emodin, enniatin B, enniatin B1, neoechinulin A) were identified in dust samples. Mould secondary metabolites can be a contamination of the litter, and feed coming from other sources (e.g., walls with mould growth on premises for poultry breeding that with air movements can get into the dust. Secondary metabolites are stable in the environment and they can be detected after mould death.

Table 8. Secondary metabolites in settled dust samples.

Secondary Metabolite	Concentration (µg/kg)									
	I	II	III	IV	V	VI	VII	VIII	IX	X
15-Hydroxyculmorin	-	-	-	-	-	-	-	114.58	-	-
3-Nitropropionic acid	1.69	3.42	1.64	1.67	2.70	-	2.48	-	-	-
α-Zearalenol	-	-	-	-	-	-	-	36.93	-	-
Alternariol	-	-	-	-	-	-	-	2.42	2.53	-
Altersetin	-	-	-	-	-	-	-	11.51	13.36	9.83
Apicidin								3.18	-	2.87
Asperglaucide	5.21	38.26	0.39	0.05	32.68	31.30	13.39	45.23	46.24	51.09
Aurofusarin	2.96	3.56	-	10.46	15.07	9.79	14.96	281.44	-	-
Beauvericin	-	-	-	-	-	-	-	1.25	1.94	-
β-Zearalenol	-	-	-	-	-	-	-	8.14	-	-
Brevianamid F	3.56	3.06	1.26	1.99	3.58	2.89	1.18	63.15	32.67	65.95
Citreorosein	-	-	-	-	-	-	-	-	13.18	-
Deoxynivalenol	-	-	-	-	-	-	-	151.20	-	-
DON-3-glucoside	2.91	-	-	-	-	2.22	3.76	-	-	-
Emodin	-	-	0.22	-	0.28	0.26	0.41	7.15	17.41	2.88
Enniatin B	0.09	0.12	0.15	0.13	0.31	0.95	0.12	2.13	0.67	0.70
Enniatin B1	0.09	0.12	0.11	0.11	0.28	1.30	0.12	1.96	0.94	0.95
Equisetin	-	-	-	-	-	-	-	1.92	1.84	2.37
Infectopyron	-	-	-	-	15.87	21.52	7.62	249.12	125.49	122.26
Mycophenolic acid	0.26	4.13	-	-	-	-	-	-	-	-
Neoechinulin A	1.73	4.26	0.14	-	0.94	0.26	0.65	22.93	426.40	784.48
Nivalenol	-	-	-	-	-	-	-	-	19.57	24.88
Siccanol	6.37	18.42	11.17	11.45	9.36	8.26	6.13	-	-	-
Zearalenone	1.62	0.71	1.16	1.32	3.04	2.07	7.30	72.18	-	-
Zearalenone-sulfate	-	-	-	-	-	-	-	204.48	42.72	49.52

"-"—not detected.

Settled dust cytotoxicity was tested on chicken hepatocyte cells, LMH (Figure 1). Tested dust samples showed weak cytotoxicity in the range of 9.2%–29.7%. Only in the case of three samples (III, V, VI), the cytotoxicity of dust towards chicken hepatocytes was higher than 20%, but still it was not significantly different ($p < 0.05$) from other farm samples.

We can hypothesize that in poultry farms, metabolites such as mycotoxins and endotoxins may be present in the form of dust suspended in water vapour, and can be inhaled by birds and people. According to our results, settled dust in poultry farms, is not a toxic agent for chicken hepatocytes under evaluated conditions. There is very little research on the influence of settled dust on birds and human health.

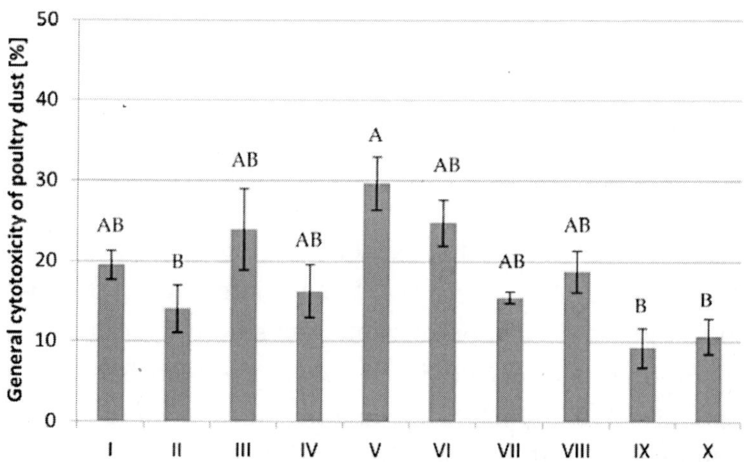

Figure 1. Cytotoxicity of poultry dust measured in MTT assay against LMH chicken cell line after 48 h exposition. Each point represents the mean of eight individual well absorbance values (± SD) for I–X poultry farm. A, B—means with the same capital letter in the same column are not significantly different (Tukey's test; $p < 0.05$).

It was also documented that volatile odorous compounds are attached to dust particles [46]. The volatile odorous compound concentrations in the air measured in poultry farms during a production cycle is presented in Table 9.

Oxygen and carbon dioxide was present at the highest concentration in all tested farms. Among volatile odorous compounds ammonia, acrolein, methyloamine, acetic acid, acetoaldehyde, and formaldehyde, were detected. We observed a variability in the concentrations of individual compounds depending on the farm and stage of the production cycle in farm II.

Analyses of odorous volatile compounds in poultry farm II were conducted before the production cycle started, and on the 20th and 35th day of breeding. Comparison of volatile odorous compound concentrations indicated that ammonia, carbon dioxide, acetaldehyde and acetic acid concentrations increased during broiler breeding cycles.

The amounts of ammonia on farms XI and XII were high at the end of the cycle: 16.8–66.7 mg/m^3. The limit according to Council Directive 2007/43/EC is 20 ppm, ~14 mg/m^3 under conditions of 1013 hPa and 293 K [23]. According to the requirements of the Council Directive 2007/43/EC [23], the maximum concentration of ammonia did not exceed permissible limits only at farm II, but the concentration of carbon dioxide exceeded the admissible value of 3000 ppm (~5490 mg/m^3 under conditions of 1013 hPa and 293 K) in the third stage of the production cycle at poultry farm II. At farm XI the concentration of carbon dioxide was very close to the limit. Assessing the impact of the substances analysed on the quality of perceived smell from poultry operations, we noted that only the concentration of acetic acid (10.02 mg/m^3) at the poultry farms (stage II) clearly exceeded the olfactory threshold (2.0 mg/m^3) [47]. However, trimethylamine, mercaptans and hydrogen sulfide were detected in small concentrations. Hence, it can be expected that the sensory perception of smell is also influenced by these substances.

Table 9. Concentration of volatile odorous compounds in the air.

No.	Compounds	Concentration (mg/m^3)				
		Farm XI	Farm XII	Farm II during Production Cycle		
				I Stage	II Stage	III Stage
1	Acetaldehyde	0.00449	0.00691	0.00437	0.00636	0.00760
2	Acetic acid	<0.11	<0.11	<0.11	10.02	0.15
3	Acrolein	<0.017	0.135	0.145	<0.017	<0.017
4	Ammonia	16.85	66.7	0.81	2.12	1.59
5	Butyl mercaptan	<0.42	<0.41	<0.43	<0.41	<0.41
6	Carbon dioxide	5464	4459	1039	4704	6564
7	Carbon monoxide	<129	<129	<132	<129	<0.13
8	Diethylamine	<0.28	<0.28	<0.29	<0.28	<0.28
9	Dimethylamine	<0.22	<0.22	<0.22	<0.22	<0.22
10	Ethyl mercaptan	<0.27	<0.29	<0.29	<0.28	<0.28
11	Ethylamine	<0.36	<0.36	<3.70	<0.36	<0.36
12	Formaldehyde	0.07067	0.07577	0.08544	0.09168	0.06900
13	Formic acid	<0.0015	<0.002	<0.002	<0.0015	<0.0015
14	Hydrogen sulfide	<0.16	<0.16	<0.16	<0.16	0.10
15	Methyl mercaptan	<0.22	<0.22	<0.23	<0.22	<0.22
16	Methylamine	<0.66	<0.66	<0.68	0.82	<0.66
17	Oxygen	286,147	305,166	316,512	304,625	249,871
18	Propyl mercaptan	<0.35	<0.35	<0.36	<0.35	<0.35
19	total organic carbon	<1.60	<1.6	<1.60	<1.60	<1.60
20	triethylamine	<0.22	<0.22	<0.22	<0.22	<0.22
21	trimethylamine	<0.13	<0.13	<0.13	<0.13	<0.13

Stages: I stage—(without chickens); II stage (20th day), III stage (35th day), XI and XII (57th week, at the end of the cycle production).

Data on threshold values concerns single chemical constituents without other chemicals present in the air. Very limited data are available on the simultaneous, combined effect of different odorous substances on the overall odour. Such an effect, caused by the mixture of substances in the exhaust air produced from poultry units, may be the source of the unpleasant smell, even though single odorants are emitted only in very small amounts.

4. CONCLUSIONS

Settled dust can be a carrier of microorganisms, odours and secondary metabolites in poultry farms. Airborne total dust concentration on poultry farms was 1.44 mg/m^3 with a high percentage of the PM_{10} fraction. Microorganisms were detected in settled dust at the following levels: 3.2×10^9cfu/g for bacteria and 1.2×10^6 cfu/g for fungi. Potential pathogens (*Enterococcus* spp., *Escherichia coli*, *Salmonella* spp., *Aspergillus fumigatus, Paecilomyces variotii*) were also found in the settled dust. Secondary metabolites detected include aurofusarin, deoxynivalenol, 15-hydroxyculmorin zearalenone, zearalenone-sulfate, infectopyron, and naochinulin A, but tested dust samples showed weak cytotoxicity towards chicken hepatocyte cells in the range of 9.2%–29.7%. Among volatile odorous compounds ammonia, acrolein, methyloamine, acetic acid, acetoaldehyde and formaldehyde were detected in the air of the tested poultry farms.

Settled dust, airborne microorganisms, mould secondary metabolites and odours detected in poultry farms working environment may result potential respiratory health implications for workers. Future research should focus on medical respiratory tests of workers taking into account all factors constituted occupational risk (physical, biological and chemical) in poultry farm working environment. Also threat prevention methods e.g., assessment of the effectiveness of Filtering Respiratory Protective Devices dedicated to poultry farm staff should be developed. Moreover, studies should be undertaken to test the effectiveness of innovative systems for the purification of air and dust, and eliminating odour from poultry farms.

ACKNOWLEDGMENTS

Studies financially supported by the National Centre for Research and Development grant no. PBS2/B8/14/2014 "Innovative deodorizing biopreparation for poultry production premises".

AUTHOR CONTRIBUTIONS

Beata Gutarowska and Janusz Hermann conceived and designed the experiments; Justyna Skóra, Katarzyna Matusiak, Piotr Wojewódzki, Adriana Nowak, Anna Ligocka, Małgorzata Okrasa, Michael Sulyok performed the experiments; Justyna Skóra, Katarzyna Matusiak, Piotr Wojewódzki, Adriana Nowak, Michael Sulyok and Beata Gutarowska analyzed the data; Piotr Wojewódzki contributed materials; Justyna Skóra and Katarzyna Matusiak wrote the paper.

CONFLICTS OF INTEREST

The authors declare no conflict of interest.

REFERENCES

1. Health and Safety Executive. Statement of Evidence Respiratory Hazards of Poultry Dust. Available online: http://www.hse.gov. uk/pubns/web40.pdf (accessed on 27 November 2015).

2. Nimmermark, S.; Lund, V.; Gustafsson, G.; Eduard, W. Ammonia, dust and bacteria in welfare-oriented systems for laying hens. *Ann. Agric. Environ. Med.* **2009**, *16*, 103–113.

3. Rimac, D.; Macan, J.; Varnai, V.M.; Vucemilo, M.; Matkovic, K.; Prester, L.; Orct, T.; Trosic, I.; Pavicic, I. Exposure to poultry dust and health effects in poultry workers: Impact of mould and mite allergens. *Int. Arch. Occup. Environ. Health* **2010**, *83*, 9–19.

4. Lugauskas, A.; Krikstaponis, A.; Sveistyte, L. Airborne fungi in industrial environments: Potential agents of respiratory diseases. *Ann. Agric. Environ. Med.* **2004**, *11*, 19–25.

5. EUR-Lex. Directive 2000/54/EC of the European Parliament and of the Council of 18 September 2000 on the Protection of Workers from Risks Related to Exposure to Biological Agents at Work. Available online: http://www.biosafety.be/PDF/2000_54.pdf (accessed on 28 January 2016).

6. Dutkiewicz, J.; Śpiewak, R.; Jabłoński, L.; Szymańska, J. *Biological Occupational Risk Factors. Classification, Exposed Occupational Groups, Measurement, Prevention*; Ad Punctum: Lublin, Poland, 2007.

7. Kirychuk, S.P.; Dosman, J.A.; Reynolds, S.J.; Willson, P.; Senthilselvan, A.; Feddes, J.J.R.; Classen, H.L.; Guenter, W. Total dust and endotoxin in poultry operations: Comparison between cage and floor housing and respiratory effects in workers. *JOEM* **2006**, *48*, 741–748.

8. Yang, Z.; Aarnink, A.J.A.; de Jong, M.C.M.; Koerkamp, P.W.G.G. Airborne microorganisms from livestock production systems and their relation to dust. *Cri. Rev. Environ. Sci. Technol.***2014**, *44*, 1071–1128.

9. Herron, S.L.; Brye, K.R.; Sharpley, A.N.; Miller, D.M.; Daniels, M.B. Nutrient composition of dust emitted from poultry broiler houses in Northwest Arkansas. *J. Environ. Prot.* **2015**, *6*, 1257–1267.

10. Lacey, J.; Dutkiewicz, J. Bioaerosols and occupational lung disease. *J. Aerosol Sci.* **1994**, *25*, 1371–1404.

11. Rusca, S.; Charrière, N.; Droz, P.O.; Oppliger, A. Effects of bioaerosol exposure on work-related symptoms among Swiss sawmill workers. *Int. Arch. Occup. Environ. Health* **2008**, *81*, 415–421.

12. Donham, K.J.; Cumro, D.; Reynolds, S.J.; Merchant, J.A. Dose-response relationships between occupational aerosol exposures and cross-shift declines of lung function in poultry workers: Recommendations for exposure limits. *J. Occup. Environ. Med.* **2000**, *42*, 260–269.

13. Kośmider, J.; Mazur-Chrzanowska, B.; Wyszyński, B. *Odory*; PWN: Warszawa, Poland, 2002; p. 180.

14. European Commission. Integrated Pollution Prevention and Control (IPPC): Reference Document on Best Available Techniques for Intensive Rearing of Poultry and Pigs. Available online: http:// eippcb.jrc.ec.europa.eu/reference/BREF/irpp_bref_0703.pdf (accessed on 28 January 2016).

15. Fournel, S.; Pelletier, F.; Godbout, S.; Lagace, R.; Feddes, J.J.R. Odour emissions, hedonic tones and ammonia emissions from three cage layer housing systems. *Biosyst. Eng.* **2012**, *112*, 181–191.

16. Nicholson, F.A.; Chambers, B.J.; Walker, A.W. Ammonia emissions from broiler litter and laying hen manure management systems. *Biosyst. Eng.* **2004**, *89*, 175–185.

17. Hayes, E.T.; Curran, T.P.; Dodd, V.A. Odour and ammonia emissions from intensive poultry units in Ireland. *Bioresour. Technol.* **2006**, *97*, 933–939.

18. Gates, R.S.; Casey, K.D.; Wheeler, E.F.; Xind, H.; Pescatore, A.J. US broiler housing ammonia emissions inventory. *Atmos. Environ.* **2008**, *42*, 3342–3350.

19. Lahav, O.; Mor, T.; Heber, A.J.; Molchanov, S.; Ramirez, J.C.; Li, C.; Broday, D.M. A new approach for minimizing ammonia emissions from poultry houses. *Water Air Soil Pollut.* **2008**, *191*, 183–197.

20. Tasistro, A.S.; Cabrera, M.L.; Ritz, C.W.; Kissel, D.E. Manipulating bedding materials and PLT™ to reduce NH_3 emissions from broiler manure. *Bioresour. Technol.* **2008**, *99*, 1952–1960.

21. Sharma, N.; Choct, M.; Wu, S.; Smillie, R.; Swick, R. Dietary composition affects odour emissions from meat chickens. *Anim. Nutr.* **2015**, *1*, 24–29.

22. EUR-Lex. Directive 2008/50/EC of the European Parliament and of the Council of 21 May 2008 on Ambient Air Quality and Cleaner Air for Europe. Available online: http://eur-lex.europa.eu/legal-content/EN/ALL/?uri=CELEX:32008L0050 (accessed on 28 January 2016).

23. EUR-Lex. Council Directive 2007/43/EC of 28 June 2007 Laying down Minimum Rules for the Protection of Chickens Kept for Meat Production. Available online: http://eur-lex.europa.eu/legal-content/EN/TXT/?uri=CELEX% 3A32007 L0043 (accessed on 28 January 2016).

24. Crook, B.; Easterbrook, A.; Stagg, S. Exposure to Dust and Bioaerosols in Poultry Farming RR655. 2008. Available

online: http://www.hse.gov. uk/research/rrpdf/rr655.pdf (accessed on 23 November 2015).

25. Bensch, K.; Groenewald, J.Z.; Dijksterhuis, J.; Starink-Willemse, M.; Andersen, B.; Summerell, B.A.; Shin, H.D.; Dugan, F.M.; Schroers, H.J.; Braun, U.; *et al.* Species and ecological diversity within the *Cladosporium cladosporioides* complex (*Davidiellaceae, Capnodiales*). *Stud. Mycol.* **2010**, *67*, 1–94.

26. Frisvad, J.C.; Samson, R.A. Polyphasic taxonomy of *Penicillium* subgenus *Penicillium*. A guide to identification of food and airborne terverticillate penicillia and their mycotoxins. *Stud. Mycol.* **2004**, *49*, 1–174.

27. Klich, M.A. *Identification of Common Aspergillus Species*; Centraalbureau voor Schimmelcultures: Utrecht, The Netherlands, 2002.

28. Pitt, J.I.; Hocking, A.D. *Fungi and Food Spoilage*; Springer: North Ryde, Australia, 2009.

29. Sulyok, M.; Berthiller, F.; Krska, R.; Schuhmacher, R. Development and validation of a liquid chromatography/ tandem mass spectrometric method for the determination of 39 mycotoxins in wheat and maize. Rapid Commun. *Rapid Commun. Mass Spectrom.* **2006**, *20*, 2649–2665.

30. Sulyok, M.; Krska, R.; Schuhmacher, R. A liquid chromatography/tandem mass spectrometric multi-mycotoxin method for the quantification of 87 analytes and its application to semi-quantitative screening of moldy food samples. *Anal. Bioanal. Chem.* **2007**,*389*, 1505–1523.

31. Malachova, A.; Sulyok, M.; Beltran, E.; Berthiller, F.; Krska, R. Optimization and validation of a quantitative liquid

chromatography-tandem mass spectrometric method covering 295 bacterial and fungal metabolites including all relevant mycotoxins in four model food matrices. *J. Chromatogr. A* **2014**, *1362*, 145–156.

32. EUR-Lex. Commission Decision of 12 August 2002 Implementing Council Directive 96/23/EC Concerning the Performance of Analytical Methods and the Interpretation of Results (2002/657/EC). Available online: http://eur-lex.europa.eu/legal-content/EN/ALL/?uri=CELEX% 3A32002 D0657 (accessed on 28 January 2016).

33. Ellen, H.H.; Bottcher, R.W.; von Wachenfelt, E.; Takay, H. Dust levels and control methods in poultry houses. *J. Agric. Saf. Health* **2000**, *6*, 275–282.

34. Viegas, S.; Faísca, V.M.; Dias, H.; Clérigo, A.; Carolino, E.; Viegas, C. Occupational exposure to poultry dust and effects on the respiratory system in workers. *J. Toxicol. Environ. Health Sci. A* **2013**, *76*, 230–239.

35. Kulkarni, P.; Baron, P.A.; Willeke, K. *Aerosol Measurement: Principles, Techniques, and Applications*; John Wiley and Sons Inc.: New York, NY, USA, 2011.

36. Health and Safety Executive. Control of Substances Hazardous to Health: The Control of Substances Hazardous to Health Regulations 2002 (as Amended). Available onlinehttp:// www.hseni.gov.uk/l5_ control_of_ substances_ hazardous_to_ health.pdf (accessed on 27 November 2015).

37. Barker, K.J.; Purswell, J.L.; Davis, J.D.; Parker, H.M.; Kidd, M.T.; McDaniel, C.D.; Kiess, A.S. Distribution of bacteria at different poultry litter depths. *Int. J. Poult. Sci.* **2010**, *9*, 10–13.

38. Witkowska, D.; Chorąży, Ł.; Mituniewicz, T.; Makowski, W. *Microbiological Contamination of Bedding and Air during the Rearing of Broilers*; Institute of Technology and Life Sciences: Falenty, Poland, 2010; pp. 201–210.

39. Lee, S.A.; Adhikari, A.; Grinshpun, S.A.; McKay, R.; Shukla, R.; Reponen, T. Personal exposure to airborne dust and microorganisms in agricultural environments. *J. Occup. Environ. Hyg.* **2006**, *3*, 118–130.

40. De Hoog, G.S. Risk assessment of fungi reported from humans and animals. *Mycoses* **1996**,*39*, 407–417.

41. David, B.; Moe, R.O.; Michel, V.; Lund, V.; Mejdell, S. Air quality in alternative housing systems may have an impact on laying hen welfare. Part I—Dust. *Animals* **2015**, *5*, 495–511.

42. Girgis, G.N.; Smith, T.K. Comparative aspects of Fusarium mycotoxicoses in poultry fed diets containing naturally contaminated grains. *World's Poult. Sci. J.* **2010**, *66*, 65–86.

43. Haschek, W.M.; Voss, K.A.; Beasley, V.R. Selected mycotoxins affecting animal and human health. In *Handbook of Toxicological Pathology*; Haschek, W.M., Rousseaux, E.C.G., Wallig, M.A., Eds.; Academic Press: New York, NY, USA, 2002; pp. 645–699.

44. Ben Salah-Abbès, J.; Abbès, S.; Houas, Z.; Abdel-Wahhab, M.A.; Oueslati, R. Zearalenone induces immunotoxicity in mice: Possible protective effects of radish extract (Raphanus sativus). *J. Pharm. Pharmacol.* **2008**, *60*, 761–770.

45. Gao, F.; Jiang, L.P.; Chen, M.; Geng, C.Y.; Yang, G.; Ji, F.; Zhong, L.F.; Liu, X.F. Genotoxic effects induced by zearalenone in a human embryonic kidney cell line. *Mutat. Res. Genet. Toxicol. Environ. Mutagen.* **2013**, *4*, 6–10.

46. Lacey, R.E.; Mukhtar, S.; Carey, J.B.; Ullman, J.L. A Review of literature concerning odors, ammonia, and dust from broiler production facilities: 1.Odor concentrations and emissions.*J. Appl. Poult. Res.* **2004**, *13*, 500–508.

47. Cometto-Muñiz, J.E.; Abraham, M.H. Olfactory psychometric functions for homologous 2-ketones. *Behav. Brain Res.* **2009**, *201*, 207–215.

CHAPTER 4

Mycotoxins and Mycotoxigenic Fungi in Poultry Feed for Food-Producing Animals

Mariana Vanesa Greco,[1,2] María Luisa Franchi,[3] Silvia Laura Rico Golba,[4] Alejandro Guillermo Pardo,[1,2] and Graciela Noemí Pose[1,2,3]

[1]Departamento de Ciencia y Tecnología, Universidad Nacional de Quilmes (UNQ), Roque Sáenz Peña 352, B1876BXD Bernal, Provincia de Buenos Aires, Argentina

[2]Consejo Nacional de Investigaciones Científicas y Técnicas (CONICET), Avenida Rivadavia 1917, C1033AAJ CABA, Argentina

[3]Universidad Nacional de Río Negro (UNRN), Mitre 331, 8336 Villa Regina, Río Negro, Argentina

[4]Facultad de Ciencias Veterinarias, Universidad Nacional de La Plata, Calle 60 y 118 S/N, 1900 La Plata, Buenos Aires, Argentina

ABSTRACT

Moulds are capable of reducing the nutritional value of feedstuff as well as elaborating several mycotoxins. Mycotoxin-contaminated feed has adverse effects on animal health and productivity. Also, mycotoxins may be carried over into meat and eggs when poultry are fed with contaminated feed. In a point prevalence study feedstuff used for poultry nutrition in Argentina was analyzed for fungal flora, natural incidence of selected mycotoxins, and nutritional quality. Ten mould genera were recovered, six of them known to be mycotoxigenic. More than 28 species were determined. Fumonisins were detected in all the samples (median 1,750 ppb). Forty-four out of 49 samples (90%) were contaminated with DON (median 222 ppb) and OTA (median 5 ppb). Also, 44 out of 49 samples were contaminated with aflatoxins (median 2.685 ppb), 42 samples (86%) with ZEA (median 50 ppb), and 38 samples (78%) with T2-toxin (median 50 ppb). Ninety percent of the samples had at least one type of nutritional deficiency. This study indicates the need for continuous assessment of the mycological status of animal feed production, in order to feed animals for optimal performance ensuring food safety.

1. INTRODUCTION

The rapid growth in the meat sector has been underpinned by the rising demand of poultry meat, which has consistently increased about threefold the rate of population growth over each of the past five decades [1].

The presence of microscopic fungi affects the quality of feeds, their organoleptic attributes, and nutritional quality [2]. Moulds like other microorganisms will assimilate and utilize the most readily available nutrients in the materials they grow upon and spoilage may result in the loss of 5 to 100% of the nutrients in the feed [3]. Regarding nutritional quality, lipids, proteins, and minerals are of essential importance for the proper development and growth of farm animals. The quantity and the nutritional requirements of feed depend on the weight and age of the poultry as well as the season. Healthy poultry require sufficient amounts of carbohydrates, lipids, and proteins, along with the necessary vitamins and dietary minerals [4, 5].

In addition to their negative impact on nutritional and organoleptic properties, moulds can synthesize different mycotoxins [6]. It is well known that contamination of animal feed with mycotoxins may induce sanitary disturbances and mortality among animals and secondary contamination of human consumers via eggs, meat, or milk [7]. Consumption of a mycotoxin-contaminated diet may induce acute and long-term chronic toxic effects [8, 9]. With respect to humans and animals, in general terms, mycotoxins exhibit toxic actions and are characterized by carcinogenic, mutagenic, teratogenic, and estrogenic properties [2].

Most toxic species belong to the genera Aspergillus, Penicillium, Fusarium, and Alternaria [10]. According to several authors mycotoxins such as aflatoxins, zearalenone, T2-toxin, deoxynivalenol, ochratoxin A, fumonisins, and patulin can be considered the most common mycotoxins found in feed and food [11–13].

Most mycotoxicoses of poultry are caused by an intake of low concentration of contaminants over a long whit the typical chronic symptoms of poor growth, poor feed efficiency, and suboptimal production. Ingestion of higher concentration however leads to acute clinical symptoms associated with specific vital organs, the immune system, and other aspects of avian physiology as well as mortality [2, 14].

For quality control the identification of the contaminating mycobiota is essential because it provides data on the potential production of its mycotoxins and is a helpful indicator to determine feed hygienic quality [15].

In the European Union the limits for several mycotoxins are regulated. The maximum content of aflatoxin B1 in poultry feed has been set in 0.02–0.05 ppm. Moreover, based on studies about feed to food mycotoxin transfer limits for ochratoxin A, deoxynivalenol, zearalenone, and fumonisins B1 and B2 in feed have been recommended [16–18].

In Argentina there is information on the fungal status including mycobiota and mycotoxins in poultry feed, but this is limited to the central region of the country, particularly Córdoba province and more recently to Entre Ríos province [19–25]. In Argentina three geographical regions can be clearly distinguished: the northern-central with grasslands, the southern (Patagonia) with tablelands, and the western with mountains.

The aim of the present study was to determine the hygienic and nutritional quality and the occurrence of selected groups of fungi and mycotoxins in poultry feeds in other important producing regions of Argentina, not previously studied. Samples were collected from the central region, particularly from Buenos Aires (the largest producing area of Argentina) and La Pampa provinces, and from Río Negro province (an increasing producing area belonging to the Patagonia region). The study includes enumeration and identification of mould genera and species, natural levels of mycotoxins such as aflatoxin, ochratoxin A (OTA), T2-toxin, fumonisin, deoxinivalenol (DON), and zearalenone (ZEA), and the evaluation of some variables related to nutritional quality (crude proteins, fats, humidity, and total ashes).

2. MATERIALS AND METHODS

2.1. Samples

A total of 49 representative samples (1-2 kg per sample) of nine commercially prepared and farm mixed poultry feeds were collected from Buenos Aires, La Pampa, and Río Negro provinces during 2010 (details about samples are shown in Table 1). All samples were homogenized and divided to obtain a 1 kg of working sample for analysis. Each sample was ground in a laboratory mill.

The main component of samples was corn followed by deactivated soybean, soybean pellet, wheat bran, sorghum, animal proteins, and grain mill products.

For mycological examination feed samples were immediately analyzed upon arrival or they were stored for 2-3 days in paper bags at room temperature (about 25°C). Feed samples intended for mycotoxin analysis were stored at −20°C.

2.2. Mycobiota Determination

The dilute plate technique was used for enumeration and isolation of fungi [10]. Ten grams of each milled feed sample was mixed with 90 mL of 0.1% peptone and shaken on a horizontal shaker for 20 minutes. Then, 0.1 mL of this dilution was inoculated on each of three different media: Dichloran Rose Bengal Chloramphenicol Agar (DRBC) to enumerate total culturable fungi, Dichloran 18% Glycerol Agar (DG18) to enumerate xerophilic fungi, and Dichloran Chloramphenicol Peptone Agar (DCPA) for selective isolation of Alternaria and Fusariumspecies. Plates were incubated at 25°C for 7 days. The DCPA plates were incubated under 12 hours of light: 12 hours of darkness photoperiod. For counting, plates containing 10–100 colonies were used and the results were expressed as colony-forming units per gram of sample (CFU/g) [10]. Individual CFU/g counts for each colony type, considered to be different, were recorded. Representative colonies of each type were transferred for subculturing onto plates with Malt Extract Agar (MEA) or Water Agar (WA), for moulds suspected to belong toAlternaria or Fusarium genera. Filamentous fungi were identified at genus level according to macro- and microscopic criteria in accordance with Samson et al. [26]. Fungal isolates were identified at species level:Penicillium and Aspergillus spp. according to Pitt and Hocking [10], Fusarium spp. according to Nelson et al. [27], Alternaria spp. according to Simmons [28], and other fungi according to Pitt and Hocking [10]. The isolation frequency (Fr) and relative density (RD) of genus/species were calculated according to Gonzalez et al. [29], Pacin et al. [30], and Saleemi et al. [31] as follows:

$$\text{Fr}(\%) = \frac{\text{number of samples with a genus or species}}{\text{total number of samples}} \times 100,$$

$$\text{RD}(\%) = \frac{\text{number of isolates of a genus or species}}{\text{total number of fungi isolated}} \times 100. \quad (1)$$

Aspergillus and Penicillium isolates were preserved on agar slants of MEA and Alternaria and Fusarium isolates on Potato Dextrose Agar (PDA) at 4°C and cryopreserved in 18% glycerol at −20°C.

Table 1: Characteristics of the samples analyzed.

Province	Type of feed	Number of samples	Sample characteristics	
Bs. As.	Broiler Chicken Starter	1	Commercial	Pelleted
	Broiler Chicken Starter	1	Noncommercial, private use	Milled
	Pullet Starter	4	Commercial	Pelleted
	Pullet Finisher	4	Commercial	Pelleted
	Laying Hen Finisher	3	Commercial	Pelleted
	Choique	5	Commercial	Pelleted
La Pampa	Pullet Starter	4	Commercial	Pelleted
	Pullet Finisher	4	Commercial	Pelleted
	Laying Hen Finisher	2	Commercial	Pelleted
Río Negro	Broiler Chicken Pre Starter	1	Commercial	Milled
	Broiler Chicken Starter	2	Commercial	Milled
	Broiler Chicken Starter	3	Noncommercial, private use	Milled
	Broiler Chicken Finisher	1	Commercial	Pelleted
	Broiler Chicken Finisher	3	Noncommercial, private use	Milled
	Pullet Starter	1	Commercial	Pelleted
	Pullet Finisher	1	Commercial	Pelleted
	Laying Hen Starter	1	Commercial	Pelleted
	Laying Hen Starter	1	Noncommercial, private use	Milled
	Laying Hen Finisher	1	Commercial	Milled
	Laying Hen Finisher	3	Noncommercial, private use	Milled

2.3. Mycotoxin Analysis

To evaluate mycotoxin occurrence, feed samples were subjected to quantitative analysis using ELISA-based analytical test kits for aflatoxin, ochratoxin A, T2-toxin, fumonisin, deoxynivalenol, and zearalenone (RIDASCREEN FAST, R-BiopharmAG). The extraction procedures were according to manufacturer's protocol. In brief, 5 g of ground sample was extracted with 25 mL of 70% methanol for aflatoxins, T2-toxin, ZEA, and fumonisins. For OTA and DON, samples were extracted with 12.5 mL of 70% methanol or 100 mL of distilled water, respectively. Afterwards, samples were shaken vigorously for 3 minutes and the extracts filtered through Whatman N°1 paper. Then, aflatoxins, OTA, T2-toxin, and ZEA filtrates were diluted with distilled water in the ratio 1 : 1 and fumonisin filtrates in the ratio 1 : 14. Fifty μL of the diluted filtrate per well was used for testing [32].

2.4. Nutritional Analysis

Feed samples were subjected to proximate analysis in accordance with standard methods described by the AOAC [33]. In terms of nutritional composition crude proteins (Kjeldahl) (Method 984.13), fats (Soxhlet) (Method 945.16), humidity (Method 934.01), and total ashes (to constant weight) (Method 942.05) were determined. Prior to analysis feed samples supplied as pellets were ground with a blade grinder or laboratory mill. All analyses were done in duplicate.

To obtain the moisture content the oven-drying method was used. Samples were dried overnight in an oven at 100°C, cooled in a desiccator to room temperature, and weighed. This procedure was performed as many times as necessary until constant weight. The result was obtained by weight difference with the sample at the beginning of the procedure.

All other analyses were performed on dry matter.

Determination of total ashes (mineral content) was performed by ignition of samples in a muffle oven set at 600°C until complete calcination of organic components. This is achieved when ashes obtained are white or light gray and the weight is constant. Samples were cooled in a desiccator at room temperature and weighed. At the beginning of this procedure samples were heated until obtention carbonaceous residue.

A Soxhlet extractor was used to determine free lipid content. Dried samples were weighed into an extraction thimble. The extraction was performed using petroleum ether of 60–80 boiling range for approximately 2 hours (occurring at least 7 cycles of filling and siphoning of the extraction tube). Then the flask with the solvent-free lipid extract was dried in an oven, cooled in a desiccator, and weighed. Results are expressed as percentage of lipids.

Determination of total nitrogen (crude protein) was conducted using the Kjeldahl method. The percentage of crude protein was calculated by multiplying the total nitrogen by a correction factor of 6.25.

2.5. Statistical Analysis

Fungal counts and toxins content were analyzed with the Kruskal-Wallis test. Multiple Range Test for variables was employed to compare means of fungal counts and toxins content of samples from different provinces. Differences were considered significant when. Statgraphics Centurion XVI version 16.1.18 was used for all analyses.

3. RESULTS AND DISCUSSION

Fungal contamination of animal feed, with the consequent mycotoxin production, is one of the major threats to human and animal health [34]. In our study, all the samples analyzed were contaminated with fungi and mycotoxins.

Fungal counts (CFU/g) were obtained from different poultry feed samples on the three culture media and the mycological quality was determined according to Gimeno [35]: samples can be qualified as good (count range < 3.10^4 UFC/g), regular (count range 3.10^4–7.10^4 UFC/g), and bad (>7.10^4 UFC/g). On both media, DRBC and DCPA, fungal count range was 10–10^6 CFU/g. According to the mycological quality criterion 56% (26/46) samples could be qualified as good, 7% (3/46) as regular, and 37% (17/46) as bad (on DRBC). In Argentina there is information about the mycobiota and natural occurrence of mycotoxins in feedstuff [19–24, 32, 36, 37]. However these reports on poultry feed refer mostly to Córdoba and Entre Rios provinces (central area of the country). With respect to the hygienic quality of poultry feed the last report from Córdoba province [24] was informed that 79% of finisher poultry feed samples exceeded the maximum allowed total fungal count to ensure hygienic quality. In Entre Rios province, 120 pelleted poultry feed samples were analyzed and total fungal count was below the established value (1×10^4 UFC/g) [25]. Significant differences among fungal contamination levels in feed from Río Negro province were observed () while nonsignificant differences were found between feed samples from each province. These results indicate the need for continuous assessment of the mycological status of animal feed production.

Xerophilic fungal counts were in the range 1×10^2–1.45×10^6 CFU/g. Spoilage of feedstuff can be due to xerophilic fungi, which are capable of rapid growth above about 0.77 water activity () and of slow growth at 0.75 and below-down to about 0.68 [10].

Although of poultry feed is low (average 0.537) it should be kept in a dry and fresh place. If by carelessness or negligence feed is not kept under proper conditions, as the humidity increases fungal growth and proliferation of xerophilic fungi can occur. Fungal counts average and median were similar between all culture media tested (10^5 CFU/g and 10^3 CFU/g, resp., data not shown).

The mycobiota and natural incidence of aflatoxin B1, ZEA, and DON were investigated by Dalcero and colleagues [19, 20]. In the first study Penicillium (98%), Fusarium (87%), and Aspergillus (52%) were the most prevalent genera determined whereas in the second study Aspergillus (85%) and Fusarium (70%) were the dominant genera. The predominant species of the genera abovementioned were particularly studied [21, 22]. The predominant species of Aspergillus were A. flavus and A. parasiticus. For Penicillium they were P. brevicompactum, P. purpurogenum, and P. oxalicum. With regard to Fusarium genus the predominant species were F. moniliforme and F. nygamai. Aflatoxin B1 was the most prevalent mycotoxin. The last study [24] indicates that A. flavus was the only species belonging to section Flavi which was isolated while Fusarium verticillioides was the prevalent species. In Entre Rios province A. flavus and A. parasiticus were the aflatoxigenic species isolated.

In our study ten mould genera were recovered, four of them known to be mycotoxigenic [10]. The frequency (Fr) and relative density (RD) are showed in Table 2. While the predominant genera are coincident with the abovementioned results, the most frequent mycotoxigenic fungi were those from the genus Fusarium (69.6), followed by Eurotium (52.2%), Penicillium (45.65%), and Aspergillus (43.5%), although the presence ofEurotium spp. was not previously reported. The occurrence of Aspergillus spp. and Fusarium spp. is widespread in tropical countries, also, in corn in Argentina [38]. Corn is the main component of poultry feed in the samples analyzed.

Table 2. Fungal genus present in poultry feed samples.

Genus	Number of isolates	Fr (%)*	RD (%)**
Aspergillus	20	43.5	10.5
Cladosporium	22	47.8	11.5
Eurotium	24	52.2	12.5
Fusarium	32	69.6	16.7
Mucor	18	39.1	9.4
Penicillium	21	45.6	10.9
Paecilomyces	1	2.2	0.5
Scopulariopsis	2	4.3	1.0
Others	16	34.8	8.4
Yeast	36	78.3	18.8

*Isolation frequency.
**Isolation relative density.

More than 28 species were recovered from poultry feed samples. Several species from different genera were determined in each sample. Up to 12 different species were recovered in several samples. The frequency (Fr) and relative density (RD) are showed in Table 3. Different to the studies from Cordoba province, the most prevalent mycotoxigenic Fusarium species determined was F. proliferatum (Fr 69.5%–RD 17.7%). In accordance with previous studies in Córdoba [19, 20] and Entre Rios provinces [25], A. flavus and A. parasiticus (Fr 6.5%) were the aflatoxigenic species isolated. Penicillium chrysogenum and P. nalgiovense (Fr 13.0%) were the most prevalent Penicillium species recovered. Some other studies have shown that most poultry feeds are contaminated with species belonging to Penicillium, Fusarium, and Aspergillus genera [23]. From 120 samples of poultry feed Magnoli et al. [23] reported the presence of 15 genera of filamentous fungi. Fusarium andPenicillium were isolated in 67.5% of the samples and Aspergillus in 57.5% of them [23]. In South America, mycobiota and natural occurrence of aflatoxins, fumonisins, and ZEA in poultry feed were studied in Rio de Janeiro, Brazil [39]. The most frequent genus isolated was Penicillium (41.26%) followed by Aspergillus (33.33%) and Fusarium (20.63%). Penicillium citrinum, A. flavus F. verticillioides, and F. graminearum were the most prevalent species in decreasing order. On the other hand, we found Fusarium and Penicillium species, followed by Aspergillus, as prevalent mycobiota (Tables 2 and 3). Eurotium species such as E. amstelodami, E. repens (Fr 26.1%), E. chevalieri (Fr 19.6%), and E. rubrum (15.2%) were also found.

Table 4 shows the distribution concentration of mycotoxins in the poultry feed tested in this study. Fumonisins were detected in all the samples analyzed in a range of 222–6,000 ppb (median 1,750 ppb). Forty-four out of 49 samples (90%) were contaminated with DON (median 222 ppb) and OTA (median 5 ppb). Also, 44 out of 49 samples were contaminated with aflatoxins (median 2.685 ppb), 42 samples (86%) were contaminated with ZEA (median 50 ppb), and 38 samples (78%) were contaminated with T2-toxin (median 50 ppb). Fumonisins and DON showed the highest levels detected (6,000 ppb and 318 ppb, resp.). Significant differences were found for fumonisins in feed from La Pampa, for OTA and T2-toxin in feed from Río Negro, and for ZEA in feed from Buenos Aires. However, no differences were found for aflatoxins ($P = 0.0604$) and DON ($P = 0.1211$).

Magnoli et al. [23] also found that fumonisins had the highest incidence and were found in 97% of the analyzed samples followed by aflatoxin B1 (46%), ZEA (18%), and DON (6%). Jaramillo [40] found that fumonisin B1 and ZEA occurred in most of the samples, whereas aflatoxin B1 occurred in lower proportion. Aflatoxin B1 values ranged between 1.2 and 17.5 ppb. Fumonisin B1 levels ranged between 1,500 and 5,500 ppb. Zearalenone levels ranged between 100 and 7,000 ppb. In both reports, the simultaneous occurrence of two carcinogenic mycotoxins, aflatoxin B1 and fumonisin B1, was determined in feed intended for poultry consumption. In Córdoba province several mycotoxins were monitored in feeds. One hundred percent of poultry samples were contaminated with fumonisins B1, but aflatoxin B1 and T2- toxin, in between other selected mycotoxins, were not detected in any poultry feed [24]. On the other hand, cooccurrence of fumonisin B1, T2-toxin, and HT-2 was detected in 100% of the samples of poultry feed analyzed in Entre Rios province, with mean levels from 4502 to 5813, 19.6 to 30.3, and 6.7 to 21.6 ppb, respectively. Also, samples were cocontaminated with aflatoxin B1 [25].

In this work cooccurrence of mycotoxins was determined in all of the samples analyzed. Cooccurrence of at least three out of six mycotoxins was determined in all of the samples analyzed. Cooccurrence of six mycotoxins was determined in 57% (28/49) of the samples. The combined action of mycotoxins can generate an interactive effect such as additivity, synergism, or antagonism. The synergistic interaction causes the most toxic effects in the case of aflatoxins and OTA, or aflatoxins and toxin T2. An additive effect of aflatoxins and DON in fattening poultry has been demonstrated. This effect was obvious in proventriculus weight, blood glucose level, and lactate dehidrogenase activity. Synergistic effects in fattening poultry were observed in the case of aflatoxins and OTA. This interaction provoked reduction of body weight and increased mortality. Moreover, synergistic effects were observed for aflatoxins and toxin T2. A reduction in body weight, increased relative weight of kidneys, gizzard and hearth, and reduction of the mean corpuscular volume and potassium levels in blood were observed [41]. In our study cooccurrence of aflatoxins and OTA was observed in 42 out of 49 samples. The cooccurrence of aflatoxins and T2-toxin was observed in 37 samples. Synergism between aflatoxins and toxin T2 is extremely important in poultry due to the prevalence and severity of both mycotoxins [40]. Simultaneous occurrence of two carcinogenic mycotoxins, aflatoxins and fumonisins, was recorded in 44/49 (90%) of the samples.

Table 3. Fungal species present in poultry feed samples.

Species	Number of isolates	Fr (%)*	RD (%)**
Eurotium amstelodami	12	26.1	6.6
E. chevalieri	9	19.6	4.9
E. herbariorum	1	2.2	0.5
E. repens	12	26.1	6.6
E. rubrum	7	15.2	3.9
Eurotium sp.	2	4.3	1.1
Aspergillus candidus	4	8.7	2.2
A. flavus	3	6.5	1.7
A. fumigatus	1	2.2	0.5
A. niger	3	6.5	1.7
A. niveus	1	2.2	0.5
A. parasiticus	3	6.5	1.7
A. terreus	2	4.3	1.1
A. versicolor	1	2.1	0.5
Aspergillus sp.	7	15.2	3.9
Cladosporium cladosporioides	22	47.8	12.1
Penicillium brevicompactum	1	2.2	0.5
P. chrysogenum	6	13.0	3.3
P. citrinum	1	2.2	0.5
P. decumbens	1	2.2	0.5
P. funiculosum	3	6.5	1.7
P. implicatum	3	6.5	1.7
P. nalgiovense	6	13.0	3.3
P. oxalicum	2	4.3	1.1
P. pinophilum	2	4.3	1.1
P. variabile	1	2.2	0.5
Penicillium sp.	6	13.0	3.3
Fusarium proliferatum	32	69.5	17.7
Fusarium sp.	5	10.9	2.8
Paecilomyces variotii	1	2.2	0.5
Scopulariopsis brevicaulis	1	2.2	0.5
S. fusca	1	2.2	0.5
Scopulariopsis sp.	1	2.2	0.5
Mucor sp.	18	39.1	9.9

*Isolation frequency.
**Isolation relative density.

Table 4. Concentration of different mycotoxins in poultry feed samples (ND: not detected).

Province	Parameter	Mycotoxins (ppb)					
		Aflatoxin*	DON*	Fumonisin*	OTA*	T2-toxin*	ZEA*
Bs. As.	Number of samples positive/tested	17/18 (94.5%)	18/18 (100%)	18/18 (100%)	17/18 (94.5%)	15/18 (83.4%)	12/18 (66.7%)
	Minimum	ND	222.00	222.00	ND	ND	ND
	Maximum	11.85	318.00	4980.00	12.31	50.00	50.00
	Median	4.32	222.00	890.00	5.00	50.00	50.00
La Pampa	Number of samples positive/tested	8/10 (80%)	8/10 (80%)	10/10 (100%)	9/10 (90%)	4/10 (40%)	9/10 (90%)
	Minimum	ND	ND	1160.00	ND	ND	ND
	Maximum	4.24	247.00	6000.00	5.00	50.00	95.58
	Median	1.96	222.00	4565.00	5.00	ND	50.00
Río Negro	Number of samples positive/tested	17/18 (94.5%)	18/18 (100%)	18/18 (100%)	18/18 (100%)	18/18 (100%)	18/18 (100%)
	Minimum	ND	222.00	222.00	5.00	ND	50.00
	Maximum	37.67	222.00	6000.00	13.20	107.23	75.76
	Median	2.47	222.00	708.50	5.00	50.00	50.00

*LOD (ppb): aflatoxin: <1.7; DON: 200; fumonisins: 222; OTA: 5; T2-toxin: <20; ZEA: 17–41.

Around the world other researchers have studied the mycobiota and mycotoxins in poultry feed during the last years. Okoli et al. [3] analyzed the mycobiota of commercial poultry feed in Nigeria. The common moulds isolated were Aspergillus spp., Penicillium spp., Mucor spp., Rhizopus spp., Epicoccum spp., Gymnoascus spp., Cladosporium spp., Mortierella spp., yeast, and bacteria. In Pakistan, Saleemi et al. [31] found Aspergillusspecies as most predominant fungi, followed by Penicillium, Fusarium, and Alternaria. Among the Aspergillusisolates, A. niger (37.74%) was the most frequently isolated species followed by A. flavus (22.64), A. ochraceous(16.98%), A. parasiticus (13.21%), A. carbonarius (3.77%), A. fumigatus (3.77%), and A. oryzae (1.89%). The proportion of toxigenic fungi among Aspergillus isolates was 73.58%. Shareef [6] recovered fourteen mould genera from Iraq poultry feed. The most frequent fungi were Aspergillus followed by Penicillium, Mucor, Rhizopus, Scopulariopsis, Alternaria, and Eurotium. They also study the presence of 4 mycotoxins; aflatoxins and ochratoxins were determined in 91.1% of the samples, fumonisins were determined in 51.1%, and T2-toxin was determined in 2.2% of the samples [3, 6, 31].

The toxicogenic fungal contamination of the raw materials occurs during the preharvest and/or the postharvest periods, and the finished feeds are exposed during production, processing, transportation, and storage [24]. It has been observed that moulds and mycotoxins presence varies depending on the geographical location and the year. Temperature and humidity play important roles not only in the development of fungi but also in mycotoxin production [24].

With respect to nutritional aspects, poor health can be caused by an imbalance of nutrients either by deficiency or by excess. Also, malnutrition if prolonged eventually leads to animal death. We found that 90% of the analyzed samples presented some kind of nutritional deficiency. The results are showed in Table 5. Forty-one out of 42 samples contained adequate moisture. However, 90% of them had at least one type of nutritional deficiency. Nineteen out of 38 samples did not reach the minimum of protein content required, 31 out of 40 did not contain the minimum requirements of lipids, and 9 out of 38 exceeded the ash content.

Table 5. Nutritional composition of poultry feed samples analyzed and recommendations (NA: not analyzed).

Animal/type of animal feed	Sample	Nutritional analysis			
		Moisture	Proteins	Lipids	Ashes
Broiler Chicken Prestarter	P33	10,29	21,34	3,65	5,89
Recommendations:	Protein (min) = 22% Lipids (min) = 5–5,5% Ashes (max) = 6–8% Moisture (max) = 12%				
Broiler Chicken Starter	P1	12,88	NA	NA	NA
	P18	11,75	NA	NA	NA
	P32	8,98	22,99	1,78	6,62
	P34	9,08	24,12	2,48	6,99
	P35	12,04	19,85	2,49	5,26
	P42	Error	23,66	5,31	5,44
	P45	10,73	21,13	4,55	6,15
Recommendations:	Protein (min) = 20–22% Lipids (min) = 3–5% Ashes (max) = 4.9–8% Moisture (max) = 12-13%				
Pullet Starter	P12	10,21	22,21	2,89	6,32
	P8	10,20	10,58	1,92	9,67
	P13	9,48	19,56	4,39	11,45
	P16	9,10	10,46	1,79	10,63
	P19	Error	21,11	4,97	5,46
	P21	NA	NA	NA	NA
	P24	11,00	NA	2,38	7,97
	P28	10,35	21,48	2,68	5,80
	P39	9,99	18,82	4,94	5,01
Recommendations:	Protein (min) = 20% Lipids (min) = 6% Ashes (max) = — Moisture (max) = 12.5%				
Broiler Chicken Finisher	P29	11,54	NA	NA	NA
	P36	12,75	17,88	2,08	4,78
	P38	9,47	19,22	2,66	6,29
	P43	11,63	16,93	4,30	5,38
Recommendations:	Protein (min) = 18–18.5% Lipids (min) = 4-5% Ashes (max) = 7.5–8% Moisture (max) = 12%				
Pullet Finisher	P9	7,61	11,60	4,26	NA
	P10	7,58	18,54	8,08	8,64
	P15	9,64	19,46	8,90	5,66
	P17	7,44	10,78	3,61	10,20
	P20	10,61	20,81	4,99	6,41
	P22	10,18	18,35	6,16	8,50
	P27	7,37	20,26	8,59	6,53
	P25	10,77	NA	2,70	NA
	P40	9,44	19,03	5,27	4,74

Recommendations:	Protein (min) = 19.5% Lipids (min) = 5% Ashes (max) = — Moisture (max) = 12.5%				
Laying Hen Starter	P31	8,94	19,06	3,25	17,28
	P46	11,85	20,72	4,69	9,03
Recommendations:	Protein (min.) = 20% Lipids (min) = 6% Ashes (max) = 5% Moisture (max) = 12%				
Laying Hen Finisher	P7	8,63	14,10	3,25	11,29
	P11	8,08	13,36	3,21	15,89
	P14	8,86	18,86	3,25	12,42
	P23	10,01	13,38	3,89	14,00
	P26	NA	NA	NA	NA
	P30	10,18	NA	NA	NA
	P37	11,34	18,82	2,33	10,37
	P41	11,34	18,82	2,33	10,37
	P44	8,83	12,21	3,91	9,51
Recommendations:	Protein (min) = 16% Lipids (min) = 6% Ashes (max) = 5% Moisture (max) = 12%				
Recommendations:	Protein (min) = 16% Lipids (min) = 6% Ashes (max) = 5% Moisture (max) = 12%				
Choique (*Pterocnemia pennata*)	P2	9,60	22,01	3,22	10,07
	P3	9,85	21,69	3,60	10,45
	P4	9,20	21,94	3,29	10,02
	P5	8,97	21,97	3,25	10,04
	P6	9,75	21,97	3,25	10,04
Recommendations:	Protein (min) = 20% Lipids (min) = 5% Ashes (max) = 11% Moisture (max) = 12%				

Recommendations values were obtained from producers.

This information is quite important considering poultry health, and hence productivity requires proper supply of these nutrients. The nutritional quality of chicken meat lies in their protein, iron, zinc, phosphorus, niacin, folic acid, and vitamins A, B6, and B12 content. The nutritional quality of the chicken thus depends on their feeding conditions.

Moulds are capable of reducing the nutritional values of feedstuff. When conditions are optimal for moulds to colonize grains in the field or in storage, their first effect is utilization of nutrients for their metabolism and propagation. This results in decreased nutritional value of feeds. Energy, crude protein, and crude fat values of mouldy corn were decreased by 5, 7, and 63%, respectively [41].

4. CONCLUSIONS

In this work, although the amounts of the mycotoxins detected on poultry feed were lower than the regulation limits established, the cooccurrence of several mycotoxins was demonstrated. The scientific literature offers a broad spectrum of information on the effects of individual mycotoxins in various animal species but concurrent exposure to multiple mycotoxins is more likely in feedstuff [8]. While mycotoxins have attracted worldwide attention due to their direct impact on human health, due to food contamination, it is also relevant to consider to what extent mycotoxins can be carried over into edible tissues like meat and eggs when poultry are fed with contaminated feed.

We suggest that periodic monitoring and the application of the Hazard Analysis and Critical Control Points (HACCP) on the prevention and control of mycotoxins in the animal feed industry are needed in order to guarantee the hygienic and nutritional quality of feed to ensure health and productivity of poultry as well as prevent human foodborne diseases.

CONFLICT OF INTERESTS

The authors declare that there is no conflict of interests regarding the publication of this paper.

ACKNOWLEDGMENTS

This work was supported by UNQ, UNRN, CONICET, and Agencia Nacional de Promoción Científica y Tecnológica (ANPCyT).

REFERENCES

1. Food and Agriculture Organization of the United Nations (FAO), Statistic Division (FAOSTAT), Trends in the livestock sector, part 3, 2013,

2. R. Cegielska-Radziejewska, K. Stuper, and T. Szablewski, "Microflora and mycotoxin contamination in poultry feed mixtures from western Poland," Annals of Agricultural and Environmental Medicine, vol. 20, no. 1, pp. 30–35, 2013.

3. I. C. Okoli, C. U. Nweke, C. G. Okoli, and M. N. Opara, "Assessment of the mycoflora of commercial poultry feeds sold in the humid tropical environment of Imo State, Nigeria," International Journal of Environmental Science and Technology, vol. 3, no. 1, pp. 9–14, 2006.

4. J. R. Gillespie and F. B. Flanders, Modern Livestock and Poultry Production, Cengage Learning, Ontario, Canada, 8th edition, 2009.

5. G. Damerow, The Chicken Encyclopedia: An Illustrated Reference, Storey, North Adams, Mass, USA, 2012.

6. A. M. Shareef, "Molds and mycotoxins in poultry feeds from farms of potential mycotoxicosis," Iraqi Journal of Veterinary Sciences, vol. 24, no. 1, pp. 17–25, 2010.

7. J. Nyamongo and M. Okioma, "The aflatoxin outbreaks in Kenya in 2004 and 2005: a case study," inProceedings of the Conference on Reducing Impact of Mycotoxins in Tropical Agriculture with Emphasis on Health and Trade in Africa, p. 3, Accra, Ghana, 2005.

8. E. M. Binder, L. M. Tan, L. J. Chin, J. Handl, and J. Richard, "Worldwide occurrence of mycotoxins in commodities, feeds and feed ingredients," Animal Feed Science and Technology, vol. 137, no. 3-4, pp. 265–282, 2007.

9. A. Venancio and R. Paterson, "The challenge of mycotoxins," in Food Safety—A Practical and Case Study Approach, A. McElhatton and R. J. Marshall, Eds., pp. 24–47, Springer, Berlin, Germany, 2007.

10. J. Pitt and A. Hocking, Fungi and Food Spoilage, Springer, Berlin, Germany, 3rd edition, 2009.

11. H. S. Hussein and J. M. Brasel, "Toxicity, metabolism, and impact of mycotoxins on humans and animals," Toxicology, vol. 167, no. 2, pp. 101–134, 2001.

12. J. L. Orellano, "Métodos de determinación, identificación y control de micotoxinas en ingredientes para la nutrición animal," Engormix, 2007.

13. A. Gimeno and M. L. Martins, Micotoxinas y Micotoxicosis en Animales y Humanos, Special Nutrients, Miami, Fla , USA, 1st edition, 2007.

14. T. Mabbett, "Keep feeds free from fungi," African Farming, pp. 15–16, 2004.

15. A. R. Khosravi, M. Dakhili, and H. Shokri, "A mycological survey on feed ingredients and mixed animal feeds in Ghom

province, Iran," Pakistan Journal of Nutrition, vol. 7, no. 1, pp. 31–34, 2008.

16. "Directive 2002/32/EC of the European Parliament and of the Council of 7 May 2002 on undesirable substances in animal feed," Official Journal of the European Communities, L 140/10, 2002.

17. Commission Regulation (EU), "Amending Annex I to Directive 2002/32/EC of the European Parliament and of the Council as regards maximum levels for nitrite, melamine, Ambrosia spp. and carry-over of certain coccidiostats and histomonostats and consolidating Annexes I and II thereto," Official Journal of the European Union, vol. L159/7, 2011.

18. "Commission recommendation of 17 August 2006 on the presence of deoxynivalenol, zearalenone, ochratoxin A, T-2 and HT-2 and fumonisins in products intended for animal feeding," Official Journal of the European Union, L229/7, 2006.

19. A. M. Dalcero, C. E. Magnoli, S. M. Chiacchiera, G. Palacios, and M. Reynoso, "Mycoflora and incidence of aflatoxin B_1, zearalenone and deoxynivalenol in poultry feeds in Argentina," Mycopathologia, vol. 137, no. 3, pp. 179–184, 1997.

20. A. M. Dalcero, C. E. Magnoli, M. Luna et al., "Mycoflora and naturally occurring mycotoxins in poultry feeds in Argentina," Mycopathologia, vol. 141, no. 1, pp. 37–43, 1998.

21. C. E. Magnoli, A. M. Dalcero, S. M. Chiacchiera, R. Miazzo, and M. A. Sáenz, "Enumeration and identification of Aspergillus group and Penicillium species in poultry feeds from Argentina,"Mycopathologia, vol. 142, no. 1, pp. 27–32, 1998.

22. C. E. Magnoli, M. A. Sáenz, S. M. Chiacchiera, and A. M. Dalcero, "Natural occurrence of Fusariumspecies and fumonisin-production by toxigenic strains isolated from poultry feeds in Argentina,"Mycopathologia, vol. 145, no. 1, pp. 35–41, 1999.

23. C. E. Magnoli, S. M. Chiacchiera, R. Miazzo et al., "The mycoflora and toxicity of feedstuffs from a production plant in Córdoba, Argentina," Mycotoxin Research, vol. 18, pp. 7–22, 2002. View at Google Scholar

24. M. del Pilar Monge, C. E. Magnoli, and S. M. Chiacchiera, "Survey of Aspergillus and Fusarium species and their mycotoxins in raw materials and poultry feeds from Córdoba, Argentina," Mycotoxin Research, vol. 28, no. 2, pp. 111–122, 2012.

25. M. P. Monge, A. M. Dalcero, C. E. Magnoli, and S. M. Chiacchiera, "Natural co-occurrence of fungi and mycotoxins in poultry feeds from Entre Ríos," Food Additives and Contaminants B, vol. 6, pp. 168–174, 2013.

26. R. A. Samson, E. S. Hoekstra, and J. C. Frisvad, Introduction to Food and Airborne Fungi, Centraalbureau voor Schimmelcultures, Utrech, The Netherlands, 7th edition, 2004.

27. P. Nelson, T. Toussoun, and W. Marasas, Fusarium Species: An Illustrated Manual for Identification, Pennsylvania State University Press, University Park, Pa, USA, 1983.

28. E. Simmons, Alternaria: An Identification Manual, American Society of Microbiology, Washington, DC, USA, 1st edition, 2009.

29. H. H. Gonzalez, S. L. Resnik, R. T. Boca, and W. F. Marasas, "Mycoflora of Argentinian corn harvested in the main

production area in 1990," Mycopathologia, vol. 130, no. 1, pp. 29–36, 1995.

30. A. M. Pacin, H. H. L. González, M. Etcheverry, S. L. Resnik, L. Vivas, and S. Espin, "Fungi associated with food and feed commodities from Ecuador," Mycopathologia, vol. 156, no. 2, pp. 87–92, 2003.

31. M. K. Saleemi, M. Z. Khan, A. Khan, and I. Javed, "Mycoflora of poultry feeds and mycotoxins producing potential of Aspergillus species," Pakistan Journal of Botany, vol. 42, no. 1, pp. 427–434, 2010.

32. M. V. Greco, A. G. Pardo, V. Ludemann, P. E. Martino, and G. N. Pose, "Mycoflora and natural incidence of selected mycotoxins in rabbit and chinchilla feeds," The Scientific World Journal, vol. 2012, Article ID 956056, 6 pages, 2012.

33. AOAC, Official Methods of Analysis, Association of Official Analytical Chemists, Gaithersburg, Md, USA, 16th edition, 1997.

34. M. D. Castillo, H. H. L. González, E. J. Martínez, A. M. Pacin, and S. L. Resnik, "Mycoflora and potential for mycotoxin production of freshly harvested black bean from the Argentinean main production area,"Mycopathologia, vol. 158, no. 1, pp. 107–112, 2004.

35. A. Gimeno, "Los Hongos y las Micotoxinas en la Alimentación Animal; Conceptos, Problemas, Control y Recomendaciones," 2002,

36. M. L. G. Pereyra, V. A. Alonso, R. Sager et al., "Fungi and selected mycotoxins from pre- and postfermented corn silage," Journal of Applied Microbiology, vol. 104, no. 4, pp. 1034–1041, 2008.

37. M. L. González Pereyra, C. M. Pereyra, M. L. Ramírez, C. A. R. Rosa, A. M. Dalcero, and L. R. Cavaglieri, "Determination of mycobiota and mycotoxins in pig feed in central Argentina," Letters in Applied Microbiology, vol. 46, no. 5, pp. 555–561, 2008.

38. G. N. Pose, Estudio de la biodiversidad de especies de Fusarium contaminantes de trigo y maíz en la provincia de Santa Fe [Ph.D. thesis], Universidad Nacional del Litoral, 2002.

39. G. R. Oliveira, J. M. Ribeiro, M. E. Fraga et al., "Mycobiota in poultry feeds and natural occurrence of aflatoxins, fumonisins and zearalenone in the Rio de Janeiro State, Brazil," Mycopathologia, vol. 162, no. 5, pp. 355–362, 2006.

40. M. Jaramillo, Aditividad , Sinergismo y Antagonismo entre Micotoxinas y sus Efectos en Pollos de Engorde, 2006,

41. A. Di Constanzo and M. Murphy, "St rategies for Feeding Mycotoxin and Mold Contaminated Grains to Cattle," 2012,

CHAPTER 5

Estimating the Nutritional Value of the Leaves of Moringaoleifera on Poultry

Addai-Mensah Donkor[1*], Richard Lander Kwame Glover[2], Daniel Addae[1], Kawku Appiah Kubi[3]

[1]Department of Applied Chemistry and Biochemistry, Faculty of Applied Sciences, University for Development Studies, Navrongo, Ghana

[2]Department of Applied Biology, Faculty of Applied Sciences, University for Development Studies, Navrongo, Ghana.

ABSTRACT

In this study, we validated the effect of Moringa oleifera leaves as food supplement on poultry. Proximate neutron activation analysis was used in determining the level of some of the vital nutritional minerals and potential toxic elements in the serum of the birds during the experimental process, including Magnesium, Copper, Vanadium, Chlorine, Aluminum, Manganese, Calcium, Mercury, Cadmium, Arsenic, Sodium and Potassium. Incorporation of Moringa oleifera leaves in maize meal fed to poultry showed profound, statistically significant effect on their weight ($p < 0.05$) as well as intensity in coloration of the beaks and legs, combs and wattles, in contrast to a control (maize-meal only) feeding of broilers ($p > 0.05$). Moreover, a significant level of the biochemical minerals of serum Calcium (1.17 - 1.20 $mmol \cdot L^{-1}$), Sodium (141.6 - 152.6 $mmol \cdot L^{-1}$), Potassium (4.1 - 5.7 $mmol \cdot L^{-1}$), Chloride (95 - 110 $mmol \cdot L^{-1}$), Albumen (30 - 55 $mmol \cdot L^{-1}$) and Triglyceride (0.3 - 1.7 $mmol \cdot L^{-1}$) were observed in broilersfedwith the formulated Moringa oleifera leaves and maize meal as compared to the maize meal only (control). Addition of Maringa oleifera leaves to poultry feed will lead to high output performance in poultry production.

KEYWORDS

Moringa oleifera; Neutron Activation; Serum; Poultry; Broilers

1. INTRODUCTION

Moringa is a tree extending in height from 5 - 12 m with an open umbrella-shaped crown, straight trunk between 10 to 30 cm thick and a corky, whitish bark. The plant, subject to climate has leaflets around 2 cm in diameter and about 2.5 cm in length. The tree yields a tuberous tap root which elucidates its acceptance to drought conditions [1]. It is primarily well-thought-out a tree of hot semi-arid regions of annual rainfall ranging from 250 to 1500 mm. Moringa is adjustable to a comprehensive array of environmental conditions from hot to humid and wet conditions. The tree is tolerant to light frosts, but does not subsist as a perennial under freezing conditions [2]. Moringa matures more speedily, reaching higher heights, when found in well-drained soils with sufficient water, but endures sandy soils, heftier clay soils and water deficient conditions. The tree can be established in marginally alkaline soils up to pH 9 as well as acidic soils as low as pH 4.5 [3] and is soundapt for a wide range of aggressive environments that would not be appropriate for other fruit, nut and tree crops [4].

Moringa oleifera Lam is one of the 14 species of the family Moringaceae, native to India, Africa, and many parts of the world [5]. Based on [6], the Moringa tree was made known to Africa from India around the twentieth century where it was to be used as a well-being complement. The plant is cited by a number of names such as horseradish tree, drumstick tree, bean oil tree, and a lot more others [7].

Exploration of the diverse parts of the plant is multidisciplinary, comprising but not restricted to nutrition, ethnobotany, medicine, analytical chemistry, phytochemistry and anthropology [8].

The Moringa plant has been expended by humans all over the century in various culinary ways [5]. Virtually all fragments of the plant are used culturally for their nutritional value, ostensible medicinal properties as well as taste and flavor. The leaves of Moringa oleifera can be consumed fresh, boiled, or kept as a dried powder for countless months supposedly without any major damage of its nutritional value [9]. Epidemiological studies have specified that Moringa oleifera leaves are a good source of nutrition and display anti-tumor, anti-inflammatory, anti-ulcer, anti-atherosclerotic and anti-convulsant activities [10] (**Table 1**). The leaves, seeds, flowers, fruit, bark and roots are all used as vegetable [11], and each part is exceptionally harvested and utilized. For example, fresh leaves are picked, shade dried, ground to a powder, and then stored for later use as a food flavoring or additive. Dried or fresh leaves are also used in foods such as soups and porridges, curry gravy and in noodles, rice or wheat [12].

Farmers have included the leaves to animal feed to support a healthy livestock [13-15] while exploiting the manure and vegetable compost for crop growth [16]. Innovative application comprise the use of Moringa powder as a fish food in aquacultural systems [17] and the Moringa leaves as a protein supplement for animals, such as cows. The feeding value of Moringa has been recounted to be analogous to that of soybeans and rapeseed meal [18]. With the leaves being rich in nutrients, pregnant women and lactating mothers use the powdered leaves to enhance their children's nourishment, principally in under developed countries where malnutrition is common [19,20].

Foods having high levels of protein (e.g., soya beans, groundnuts, sunflower seeds) and other feed supplements for poultry production are expensive to purchase while energy foods (e.g., cereal such as maize, sorghum, millets, wheat, rice, maize bran, wheat bran and rice bran by-products) are often plentiful and relatively cheap. Thishas obligated most poultry farmers to move away from this industry leading to low production of poultry meat and eggs and raised the price of poultry products due to high cost of importation. The energy in the dietnecessary for production of poultry meat and eggs and for the maintenance of vital functions and body temperature, are largely in the form of carbohydrates, fats and amino acids which can be found in the leaves of Moringa oleifera. The production of protein in the body tissues entails an adequate supply of about 20 different amino acids, 10 of which cannot be synthesized by poultry, and must therefore be provided in their diets. These essential amino acids are distributed in the leaves of Moringa oleifera all year round [21-23].

The current study sought to validate the potential vital nutritional minerals in the leaves of Moringa oleifera and their effect on the weight and other features of broilers fed with maize meal formulated with Moringa oleifera leaf powder.

2. MATERIALS AND METHOD

2.1. Materials

Moringa oleifera leaves (MOL) were harvested from an orchard at early flowering stage at Navrongo in the Kassena Nankana District of Upper East Region of Ghana. The plant material was authenticated by Dr. Walter M. Kpikpi, a botanist at the Department of Applied Biology in the Faculty of Applied Sciences of the University for Development Studies, Navrongo Campus.

4-month old broilers were purchased from the Navrongo poultry farm, while maize was obtained from the Navrongo market.

2.2. Preparation of Moringa oleifera Leaf Powder and Maize-Meal (MOLMM)

Branches were cut from the Moringa trees, threshed carefully to separate leaves from twigs and weighed. The leaves were spread out on a floor and allowed to air-dry for 4 days under shady and aerated conditions. The dried leaves were ground with mortar and pestle to obtain a leaf powder. The maize was ground at a corn mill to appreciable particle sizes. A 20 g sample of the maize powder was mixed with 5.0 g of the Moringa oleifera leaf powder to derive the MOLMM.

2.3. Management of Experimental Animals (Broilers)

Five chicken-broilers were allotted to floor pens (2:2:1), with their respective ID numbers of A1 and A2, B1 and B2, and C1) in a randomized complete block design. Sunlight was used as a source of light during the day and fluorescent tubes were used to illuminate the pens at night. The birds were kept under a deep litter management system and “saw dusts” were used as litter materials to cover the floor to serve as absorbents.

2.4. Experimental Design

A single diet was formulated in two different phases at 3-week intervals. The MOLMM was first fed to the broilers in block “A” for 21 days while those in block “B” and “C” were given only maize meal (MM). After 21 days, block “B” broilers were also given same quantities of MOLMM for 21 days while block “A” and “C” broilers were maintained on MM only. Broilers were made to acclimatize to experimental diet for a period of 7 days prior to data collection. Broilers were provided with ration twice daily at intervals of 7 hours (9:00 am-4:00 pm) and water was provided daily. The weight of each broiler was taken from each block at 3-day intervals during the period of study. Observations of color of their beaks and legs, combs and wattles as well as the activeness of the broilers were made.

2.5. Determination of Nutritional Minerals

Two milliliter blood samples were collected from the broilers in each block using new sterile needles and syringes through the veins in the wings and placed inside vacutainer test tubes labeled appropriately with corresponding treatment IDs. The blood samples were centrifuged at 3000 rpm for 10 minutes in order to separate the serum. Selected blood chemistry was carried out in the analyzer for some nutritional biochemical minerals at the War Memorial Hospital in Navrongo.

Table 1. Recommended daily intakes of the various minerals.

Minerals (nutritional nutrients)	Major function	RDI (Mg)	Deficiency	Toxicity
Calcium	Component of bone and teeth, blood clothing, nerve transmission, muscle action	800 - 1200	Osteoporosis	Kidney stone
Magnesium	Bone formation, muscle and nerve function	280 - 320	Weakness, muscle pain	Rare
Sodium	Osmotic pressure, nerve transmission	500	Nausea, vomiting, muscle cramps	Possible hypertension
Chlorine	HCl synthesis, osmotic pressure, nerve transmission	700	Rare	Possible hypertension
Potassium	Osmotic pressure, nerve transmission	2000	Heart irregularities, muscle cramps	Slowed heart rate
Copper	Iron metabolism, component of many enzymes	1.5 - 3.0	Anemia	Rare, vomiting
Manganese	Fat synthesis, bone formation	1.8 - 2.3	Poor growth	Weakness; nervous system problems; mental confusion
Vanadium	Enzymatic reactions such as metabolism of carbohydrate and lipids, glucose and protein	<1.8	Rare. Impairs glucose metabolism	Damage to both liver and kidney. Causes gastrointestinal distress, primary diarrhea

Source: USDA nutrient database.

2.6. Neutron Activation Analysis of Some Vital Nutritional Minerals in Moringa oleifera Leaf

Proximate neutron activation analysis was carried out on the Moringa oleifera leaf powder, using the Research Reactor-1 at the Ghana Atomic Energy and Commission (GAEC)to determine the presence of nutritional elements such as Magnesium, Copper, Vanadium, Chlorine, Aluminum, Manganese, Calcium, Sodium and Potassium as well as the heavy metals Mercury, Cadmium and Arsenic.

2.7. Statistical Analysis

The data were statistically dissected using one-way ANOVA. Results were considered to be statistically significant when ($P < 0.05$).

3. RESULTS

After completion of forty-two (six weeks) treatment the effects of the leaves of Moringa oleifera and maize meal on body weight and levels of some vital nutritional components in the serum of chicken-broilers as well as some nutritional content in leaves were investigated in controlled conditions.

The body weight of broilers fed with MOLMM increased significantly over the 21-day period as compared to those fed on the maize meal only diet (Figures 1 and 2). There was an increase in weight of broiler A1 initially from day 1 up to day 6, dropped up to day 9, rose from day 10 up to day 15, remained steady up to day 19 and rose till day 21 (**Figure 1**). Broiler A2 showed a similar weight increase pattern except for an initial decrease within the first three days (**Figure 1**). The weight of broiler B1 kept fluctuating over the initial 21-day period when it was fed with only MM (**Figure 2**). It increased up to day 5, dropped by same margin by day 10, stabilized up to day 15, rose and declined by same margin up to day 21 before feeding with MOLMM commenced. Its weight then sharply increased within nine days, dipped slightly for a couple of days, shot up for another five days and stabilized up to day 42. On the other hand, the weight of broiler B2 increased from day one up to three, dropped up to day 6, remained constant up to day15 and dropped up to day 21. With the change in diet to MOLMM, its weight increased almost steadily up to day 38 with two slight dips in-between, finally stabilizing till day 42 (**Figure 2**). Broiler C1 which was fed with only MM throughout the period of study started with a constant weight for the first 6 days which dropped sharply up to day 9, stabilized for the next 6 days, increased up to day 18 and decreased again till day 21. Finally, it moved up again from day 24 to 27 and remained at that level till the end of the study (**Figure 2**).

Effective yellow coloration of the beaks and legs as well as bright coloration of the combs and wattles was observed in broilers fed with MOLMM as compared with only MM-fed broilers. To determine whether or not there was a statistically significant difference achieved by Moringa oleifera leaves meal and maize meal during feeding, A significant effect ($P < 0.05$) on body weight of MOLMM-fed and insignificant ($P > 0.05$) on body weight of MM-fed broilers were observed. Higher activity of broilers fed with MOLMM as compared with those on a MM only diet was also observed.

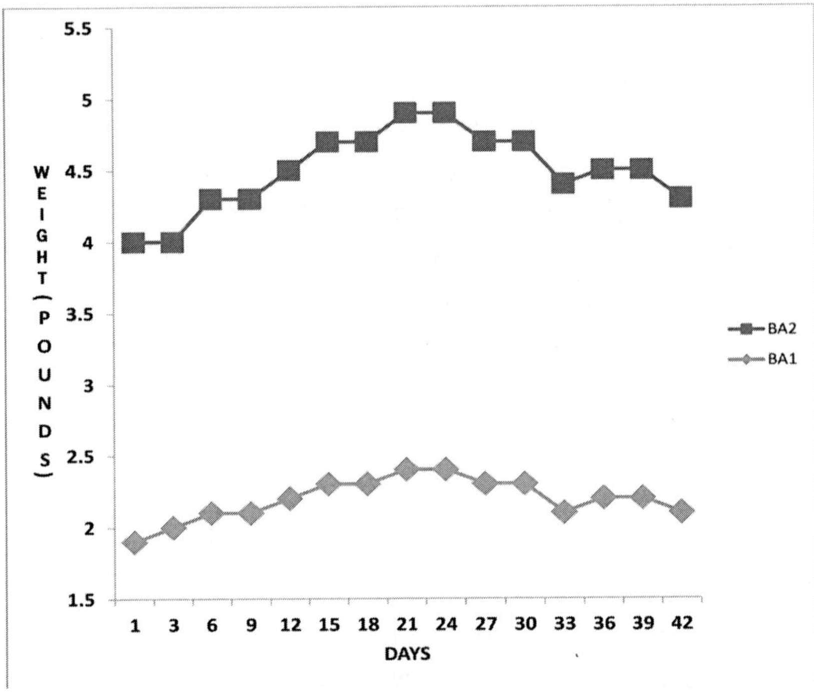

Figure 1. (■) Variations in weight of broiler A1 fed on Moringa oleifera leaf powder maize meal (MOLMM) for 21 days, and on only maize meal after 21 days until the 42nd day. (♦)Variations in weight of broiler A2 fed on Moringa oleifera leaf powder maize meal (MOLMM) in the first 21 days and on maize meal only after 21 days until the 42nd day.

Initial serum samples were taken after 21 days of feeding the broilers due to the difficulty of drawing samples at the earliest infant stage. The total levels of serum Calcium, Sodium, Potassium, Albumen, Triglyceride and Chloride of broilers fed with MOLMM and MM only as well as the reference ranges are shown in (**Table 2**). Significant increases in total serum levels of Calcium, Sodium, Potassium, Albumen and Chloride were observed for the broilers fed with the MOLMM over the feeding period. There was, however, reduction in serum Triglyceride level of the broilers when fed with the MOLMM. Generally, there was reduction in biochemical serum minerals of broilers fed with MM only (control).

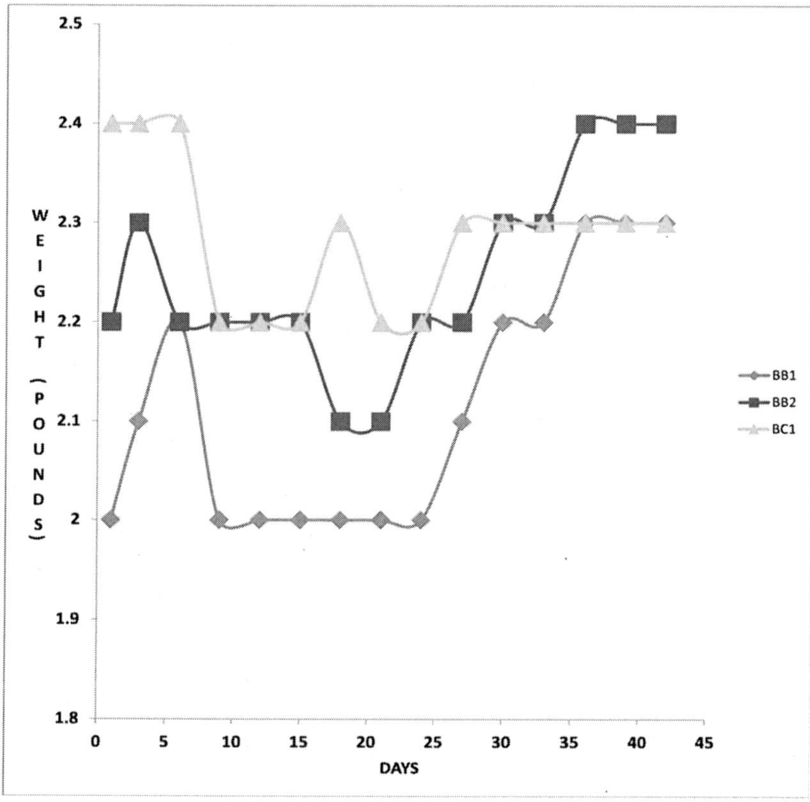

Figure 2. (♦) Variations in weight of broiler B1 fed on maize meal only for 21 days and on Moringa oleifera leaf powder maize meal (MOLMM) after 21 days until the 42nd day. (■)Variations in weight of broiler B2 fed on only maize meal for 21 days and on Moringa oleifera leaf powder maize meal (MOLMM) after 21 days until the 42nd day. (▲) Variations in weight of broiler C1 fed on only maize meal for the entire study period.

Proximate neutron activation analysis showed that Moringa oleifera leaves abound in vital nutritional minerals namely Magnesium, Copper, Vanadium, Chlorine, Aluminum, Manganese, Calcium, Sodium and Potassium. They were found to be devoid of potential toxic elements such as Mercury, Cadmium, and Arsenic (**Table 3**).

4. DISCUSSION

The present study reveals that the leaves of Moringa oleifera contain most of the major nutritional elements needed for healthy growth as revealed during neutron activation analysis of Moringa oleifera leaf (MOL) powder.

There were no detectable peaks (NDP) for the heavy metals of Mercury, Cadmium and Arsenics which can be toxic to the body when ingested; depicting that MOL is safe for consumption.

There was a significant increase in weight of broilers A1 and A2 when they were fed with MOLMM from day one up to day twenty one and remained stable up to day twenty four. There was then a significant decrease from day twenty four up to day forty two when only MM was used to feed them. The weight remained fairly constant between day twenty one and twenty four when only MM was administered. This could be as a result of the nutritional minerals which were still present in the body. It could be inferred that Moringa oleifera leaves had a significant impact on the weight of the broilers fed with MOLMM through the provision of vital biochemical minerals which might have been absent or inadequate in the normal diet of maize meal (MM) as shown for the control samples. From the present study it was revealed that MOL lowered serum Triglyceride level and increased serum Calcium, Sodium, Potassium, Albumen and Chloride. The latter is justified by the levels of these and other minerals detected in MOL powder after the neutron activation analysis.

Broilers A1 and A2 were initially fed with only maize meal for 21 days before swapping to MOLMM, whilst broilers B1 and B2 were fed with MOLMM for 21 days initially. We therefore speculate that there might be few nutritional minerals in the maize meal which might have also influence significantly the absorption of high nutritional mineral content in the MOL making broilers B1 and B2 had substantial weight gain when feed was supplemented with MOL.

The sample size selected for this particular study were low, as a result the data collected with respect to this study is preliminary and the research still on-going with current analysis comprising a large sample size in our laboratory which would be presented in our next manuscript in preparation.

5. CONCLUSIONS

Moringa oleifera leaves contain major nutritional minerals needed for healthy growth of animals including poultry. Feeding of poultry with maize meal formulated with Moringa oleifera leaf powder can lead to increase in weight, bright coloration of their combs, wattles, beaks and legs as well as general increase in activity of the birds.

Heavy metals like mercury, arsenic and cadmium which are potentially toxic are absent from the leaves of Moringa oleifera, thus making their incorporation into poultry feed safe.

Table 2. Analysis of Total Serum Calcium, Sodium, Potassium, Albumen, Triglyceride and Chloride levels of broilers at various stages of the experimental process.

Mineral	First two weeks					One month					Forty two days					Standard reference range
	A1	A2	B1	B2	C1	A1	A2	B1	B2	C1	A1	A2	B1	B2	C1	
Calcium (mmol·L^{-1})	1.01	1.61	1.03	1.05	1.00	1.05	2.01	1.51	1.32	1.60	1.31	2.21	1.78	2.00	1.40	1.20 - 1.73
Sodium (mmol·L^{-1})	138	80.0	115	128	137	157	161	136	158	149	155	138	152	155	148	141.6 - 152
Potasium (mmol·L^{-1})	2.96	4.28	4.01	3.71	5.00	3.89	4.25	3.65	5.01	4.81	21.97	4.43	6.56	1.67	4.78	4.1 - 5.7
Albumen (g·L^{-1})	32.2	50.3	41.7	40.7	48.7	58.6	38.5	59.8	50.8	50.0	48.6	44.7	59.7	55.8	48.9	30 - 55.0
Triglyceride (mmol·L^{-1})	2.32	3.10	1.81	2.81	2.23	2.15	1.00	0.34	0.70	1.68	1.03	1.30	2.85	1.83	1.62	0.30 - 1.70
Chloride (mmol·L^{-1})	97.0	101	112	96.0	91.0	117	108	98.0	120	99.0	108	101	108	104	97.0	95 - 110

Table 3. Nutritional mineral content of dried Moringa oleifera leaf powder (in ppm) and gamma energy released per nutritional mineral (in Kev).

Nutrient	Content (ppm)	Gamma energy (Kev)
Mg	3322.32	1014
Cu	2.96	1039
V	0.449	1434
Cl	1380.29	1642
Al	458.22	1779
Mn	46.07	1810
Ca	14720	3084
Hg	*NDP	77.4
Cd	*NDP	336
As	*NDP	559
Na	393.02	1369
K	13712	1524

*NDP: No Detectable Peaks.

The availability of Moringa oleifera in many rural communities in Ghana makes its use as a supplement for poultry feed an economically viable alternative to expensive commercially produced poultry feeds.

ACKNOWLEDGEMENTS

The authors are grateful to the Department of Applied Chemistry & Biochemistry, University for Development Studies, Navrongo Campus, for providing some of the materials and equipment used in this study; the Department of Applied Biology (Microbiology Laboratory) of the same university for providing the necessary materials for the biological studies; the Ghana Atomic Energy Commission (GAEC), Kwabenya, Accra for the Neutron Activation Analysis of nutritional mineral contents in dried leaf powder of Moringa oleifera and the Navrongo Health Research Centre for the analysis of total serum Calcium, Sodium, Potassium, Albumen, Triglyceride and Chloride levels of broilers at various stages of the research.

REFERENCES

1. M. L. Prince, "Maringa Tree," ECHO Technical Note, Educational Concerns for Hunger, 1985.

2. R. N. Bennett, F. A. Mellon, N. Foidl, J. H. Pratt, M. S. Dupont, L. Perkins and P. A. Kroon, "Profiling Glucosinolates and Phenolics in Vegetative and Reproductive Tissues of the Multi-Purpose Trees Moringa oleifera L. (Horseradish Tree) and Moringa stenopetala L," Journal of Agricultural and Food Chemistry, Vol. 51, 2003, pp. 3546-3553.

3. I. Oduro, W. O. Ellis and O. Deborah, "Nutritional Potential of Two Leafy Vegetables: Moringa oleifera and Ipomoea batatas Leaves," Scientific Research and Essay, Vol. 3, No. 2, 2008, pp. 57-60.

4. D. J. Kumari, "Hypoglycemic Effect of Moringa oleifera and Azadirachta indica in Type-2 Diabetes," Bioscan, Vol. 5, 2010, pp. 211-214.

5. S. Iqbal and M. I. Bhanger, "Effect of Season and Production Location on Antioxidant Activity of Moringa oleifera Leaves Grown in Pakistan," Journal of Food Composition and Analysis, Vol. 19, No. 6-7, 2006, pp. 544-551.

6. D. Mukunzi, J. Nsor-Atindana, X. M. Zhang, A. Gahungu, E. Karangwa, G. Mukamurezi, H. Al-Domi, I. L. Princewill-Ogbonna, P. C. Ogbonna and N. J. Arief, "Comparison of Volatile Profile of Moringa oleifera Leaves from Rwanda and China Using HS-SPM," Pakistan Journal of Nutrition, Vol. 10, No. 7, 2011, pp. 602-608.

7. M. S. Dawkins, P. A. Cook, M. J. Whittingham, K. A. Mansell and A. E. Harper, "What Makes Free-Range Broiler Chickens Range? In-Situ Measurement of Habitat Preference," Animal Behaviour, Vol. 66, No. 1, 2003, pp. 342-344.

8. F. Anwar, S. Latif, M. Ashraf and A. H. Gilani, "Moringa oleifera: A Food Plant with Multiple Medicinal Uses," Phytotherapy Research, Vol. 21, No. 1, 2007, pp. 17-25.

9. D. Arabshahi and B. A. Ringh, "Adoption of Moringa oleifera to Combat Under-Nutrition Viewed through the Lens of the 'Diffusion of Innovations' Theory," Ecology of Food and Nutrition, Vol. 48, No. 3, 2007, pp. 212-225.

10. P. Chumark, P. Khunawat, Y. Sanvarinda, S. Phornchirasilp, N. P. Morales, L. Phivthong-Ngam, P. Ratanachamnong, S. Srisawat and K. U. Pongrapeeporn, "The in Vitro and in Vivo Antioxidant Properties, Hypolipidaemic and Antiatherosclerotic Activities of Water Extract of Moringa

oleifera Lam. Leaves," Journal of Ethnopharmacology, Vol. 116, No. 3, 2008, pp. 439-446.

11. H. P. S. Makkar and K. Becker, "Nutritional Value and Anti-Nutritional Components of Whole and Ethanol Extracted Moringa oleifera Leaves," Animal Feed Science and Technology, Vol. 63, No. 1, 1996, pp. 211-228.

12. R. G. Abilgos and C. V. C. Barba, "Utilization of Malunggay (M. oleifera Lam.) Leaves in Rice (Orya sativa L.) Flat Noodle-Production," Philippine Journal of Science, Vol. 128, 1999, pp. 79-84.

13. J. Lako, V. C. Trenerry, M. Wahlqvist, N. Wattanapenpaiboon, S. Sotheeswaran and R. Premier, "Phytochemical Flavonols, Carotenoids and the Antioxidant Properties of a Wide Selection of Fijian Fruit, Vegetables and Other Readily Available Foods," Food Chemistry, Vol. 101, No. 4, 2007, pp. 1727-1741.

14. S. Ghasi, E. Nwobodo and J. O. Ofili, "Hypocholesterolemic Effects of Crude Extract of Leaf of Moringa oleifera Lam in High-Fat Diet Fed Wistar Rats," Journal of Ethnopharmacology, Vol. 69, No. 1, 2000, pp. 21-25.

15. A. Naznin, R. Mamunur and A. Shah, "Comparison of Moringa oleifera Leaves Extract with Atenolol on Serum Triglyceride, Serum Cholesterol, Blood glucose, Heart Weight, Body Weight in Adrenaline Induced Rats," Saudi Journal of Biological Sciences, Vol. 15, No. 2, 2008, pp. 253-258.

16. L. J. Fuglie, "The Miracle Tree, Moringa oleifera," Natural Nutrition for the Tropics, CWS, Dakar, Senegal, 1999, p. 115.

17. K. Mehta, R. Balaraman, A. H. Ami, P. A. Bafna and O. D. Gulati, "Effect of Fruits of Moringa oleifera on the Lipid

Profile of Normal and Hypercholesterolaemic Rabbits," Journal of Ethnopharmacology, Vol. 86, No. 2-3, 2003, pp. 191-195.

18. C. R. Soliva, M. Kreuzer, N. Foidl, G. Foidl, A. Machmüller and H. D. Hess, "Feeding Value of Whole and Extracted Moringa oleifera Leaves for Ruminants and Their Effects on Ruminal," Animal Feed Science and Technology, Vol. 118, No. 1, 2005, pp. 47-62.

19. P. Sudha, S. M. Asdaq, S. S. Dhamingi and G. K. Chandrakala, "Immuno-Modulatory Activity of Methanolic Leaf Extract of Moringa oleifera in Animals," Indian Journal of Physiology and Pharmacology, Vol. 54, No. 2, 2010, pp. 133-140.

20. K. Stephen, M. A. Bangert, J. William, A. Marshall and L. William, "Clinical Biochemistry Metabolic and Clinical Aspects," Elsevier, Philadelphia, 2008.

21. H. P. S. Makkar and K. Becker, "Plant Toxins and Detoxification Methods to Improve Feed Quality of Tropical Seeds," Asian-Australasian Journal of Animal Sciences, Vol. 12, No. 3, 1999, pp. 467-480.

22. S. K. Pal, P. K. Mukherjee and B. P. Saha, "Studies on the Anti-Ulcer Activity of Moringa oleifera Leaf Extract on Gastric Ulcer Models in Rats," Phytotherapy Research, Vol. 9, No. 6, 1995, pp. 463-465.

23. P. S. Kumar, D. Mishra, G. Ghosh and G. S. Panda, "Medicinal Uses and Pharmacological Properties of Moringa oleifera," International Journal of Phytomedicine, Vol. 2, 2010, pp. 210-216.

CHAPTER 6

Efficiency of Probiotics in Farm Animals

Etleva Delia[1], Myqerem Tafaj[1] and Klaus Männer[2]

[1] Faculty of Agriculture and Environment, Agricultural University of Tirana,, Albania

[2] Institut fur Tierernährung, Freie Universität Berlin, Germany

1. INTRODUCTION

The first concept of probiotics was originally developed by [38]. He suggested that ingested bacteria could have a positive influence on the normal microbial flora of the intestinal tract. Probiotics are considered as growth and health stimulators and are used extensively in animal feeding, especially in pig and poultry production.

Probiotics have been defined also by [6] as "*a live microbial feed supplement which beneficially affects the host animal by improving its intestinal balance*". There is a relatively large volume of literature that supports the use of probiotics to prevent or treat intestinal disorders. Currently, the best studied probiotics are the lactic acid bacteria, particularly *Lactobacillus sp* and *Bifidobacterium sp.*

Therefore, an intensive research work is carrying out in this topic from many researcher groups in different countries. Many years later, probiotics were determined as: viable microbial feed supplements, which are believed to stimulate growth and the health as well as to modify the ecology of the intestine in a beneficial manner for the host [3], [34], [54]. Probiotics should lead to beneficial effects for the host animal due to an improvement of the intestinal microbial balance [12] or of the properties of the indigenous micro-flora [21]. There are also many mechanisms by probiotics enhance intestinal health, including stimulation of immunity, competition for limited nutrients, inhibition of epithelial and mucosal adherence, inhibition of epithelial invasion and production of antimicrobial substances [47].

Possible modes of actions are the modification of the intestinal microorganisms and the nutrient availability with response to the morphology and histology as well as the transport physiology. Significant positive effects of probiotics on performance, health, vitality, gut ecology as well digestibility are observed in many studies, although the mode of action of probiotics is not still completely explained [24], [55], [25], [4]. Efficiency probiotic on a focus of combined preparation have hardly been concluded.

2. EFFICIENCY OF PROBIOTIC IN FARM ANIMALS

The claims made for probiotics are many and varied but it is not always possible to provide good scientific evidence to support them. However the potential benefits that can arise from applications of the probiotic concept are shown as below:

Potential beneficial effects of probiotics for farm animals by [13].

- Greater resistance to infectious diseases
- Increased growth rate
- Improved feed conversion.
- Improved digestion.
- Better absorption of nutrients
- Provision of essential nutrients
- Improved milk yield
- Improved milk quality.
- Increased egg production.
- Improved egg quality
- Improved carcass quality and less contamination

Since probiotics are discussed as alternatives to antimicrobial growth promotors their impact on performance of farm animals is of prime interest. For authorization of microorganisms as feed additives it is also required to show significant effects on performance data [54]. By far most experiments were performed with piglets. According to a literature review by [61] no significant positive effects could be found from the hitherto results with piglets and fattening pigs. Later, the evaluation of studies conducted with raising piglets drew a different picture [11]. [61] was used the strict criteria of biostatistics and only significant effects were documented. Today, trends without statistical significance are also considered as positive effect by [54]. It is obvious that majority of the experiments show trends toward positive effects, however the significance level of $p \leq 0,05$ was reached only in 5% of experiments. Due to the complexity of the intestine, individual variations of animals to probiotic inclusion may be the rule and not the exception. Considering this concept, the range between no effect and significant effects seem to be reasonable.

In a trial with 90 treated and 90 untreated *Bacillus cereus* – preparation weaned piglets; the probiotic treated animals gained 7% more live weight during 6 weeks after weaning with a reduced feed conversion ratio of 2.4%. Both results were not significant [25]. This point towards a high variation in the response of the individual animals to this type of feed additives [54].

With regard to the evaluation of animal performance, the same conclusion can be draw for experiments with fattening chicken carried out by [53].This is also reflected by a series of experiments with turkey, poultry under field conditions using three probiotics [34]. Again none of the effects in performance were significant, on average weight gain was improved by 1,5% (+0,1 to + 3,8) and feed conversion by –2% (-7 to –3,5). A further observation was a more pronounced effect of additive during weeks 1 to 5. However again no significance was seen in the period's week 1 plus 2 and 3 to 5, respectively [54].

Authors in [54] concluded that the inconsistency of the effectiveness of a feed additive is of course not convenient, but on the other hand comprehensible for this type of feed additive. Probiotic do not act like essential nutrients in term of a clear dose response until the requirements are met. Due to the complexity of intestine, individual variations of animals to probiotic inclusion may be the rule and not the exception. Considering this concept the range between no effect and significant effects seem to be reasonable.

3. MODE OF ACTION OF PROBIOTICS

The development of probiotics for farm animals is based on the knowledge that the gut microflora is involved in resistance to disease. The gut microflora has been shown to be involved in protection against a variety of pathogens including *Escherichia coli, Salmonella Camylobacter, Clostridium and Rotavirus.* Hence the probiotic approach may be effective in the prevention and therapy of these infections. No attempt will be made to summarize the evidence available for all of these effects [13].

The one area where it is possible to arrive at some scientifically based conclusions is the effect that the probiotics preparations have on resistance to infections.

The stressful conditions experienced by the young animal causes changes in the composition and/or activity of the gut microflora. Probiotic supplementation seeks to repair these deficiencies and provide the type of microflora which exists in feral animals uninfluenced by modern farm rearing methods. The products available are of varying composition and efficacy but the concept is scientifically-based and intellectually sound. Under the right conditions the claims made for probiotic preparations can be realized [13].

Table 1. Defense functions of epithelial cells [**37**].

Molecule	Defense function	References
Lysozyme	Lyses bacterial cell walls	[2], [46]
Defensins	Form pores in bacterial cell wall	[2], [42]
Mucus	Prevents bacterial adhesion made by goblet cells, a specialized epithelial cell type.	[41]
MHC class I	Presents antigen to cytotoxic T-lymphocytes	[14]
MHC Class II	Presents antigen to helper T-lymphocytes	[14]

There are many proposed mechanisms by which probiotics may protect the host from intestinal disorders. The sum of all processes by which bacteria inhibit colonization by other strains is called colonization resistance. Much work remains to classify the mechanisms of action of particular probiotics against particular pathogens. In addition, the same probiotic may inhibit different pathogens by different mechanisms. Listed below is a brief description of mechanisms by which probiotics may protect the host against intestinal disease.

Possible mode of action of intestinal bacteria can be summarized as follows by [54]:

- Increase of desired intestinal bacteria;
- Competitive adhesion to epithelial receptors;
- Production of specific substances (bacteriocins, dipicolinic acid, bioactive peptides)
- Competition for nutrients between probiotic and undesired bacteria;
- Micro-environmental pH reduction by production of acid;

- Reduction of bacterial bile salt deconjugation;
- Passive aggregation of probiotics and pathogenic bacteria;

4. PRODUCTION OF INHIBITORY SUBSTANCES

Probiotic bacteria can produce a variety of substances that are inhibitory to both gram-positive and gram-negative bacteria. These inhibitory substances include organic acids, hydrogen peroxide and bacteriocins. These compounds may reduce not only the number of viable cells but may also affect bacterial metabolism or toxin production.

5. BLOCKING OF ADHESION SITES

Competitive inhibition for bacterial adhesion sites on intestinal epithelial surfaces is another mechanism of action for probiotics [18]. Consequently, some probiotic strains have been chosen for their ability to adhere to epithelial cells. Gut bacteria prevent intestinal colonization by pathogenic organisms directly by competing more successfully for essential nutrients or for epithelial attachment sites [48].

6. COMPETITION FOR NUTRIENTS

Competition for nutrients has been proposed as a mechanism of probiotics. Probiotics may utilize nutrients otherwise consumed by pathogenic microorganisms. However, the evidence that this occurs in vivo is lacking.

7. DEGRADATION OF TOXIN RECEPTOR

The postulated mechanism by which *Sacchromyces boulardii* protects animals against *C. difficile*intestinal disease is through degradation of the toxin receptor on the intestinal mucosa [5].

8. INFLUENCE ON THE IMMUNE SYSTEM

The intestinal micro flora is an important component of host animal. A critical review of the literature indicates that probiotic supplementation of the intestinal micro flora may enhance defense, primarily by preventing colonization by pathogens and by indirect, adjuvant-like stimulation of innate and acquired immune functions [37]. The role of nonpathogenic bacteria in the development of the intestinal immune system and in protecting the host from pathogenic challenges has been studied.

Intestinal bacteria provide the host with several nutrients, including short-chain fatty acids, vitamin K, some B vitamins and amino acids [49], [67]. Intestinal bacteria also protect the host from pathogens, forming a front line of mucosal defense. The indigenous microflora induces recruitment of lamina propria immune cells, which form a second tier of defense by activation of appropriate inflammatory or immune mechanisms during infection.

Recent evidence suggests that stimulation of specific and nonspecific immunity may be another mechanism by which probiotics can protect against intestinal disease [45]. For example, per oral administration of *Lactobacillus* GG during acute rotavirus diarrhoea is associated with an enhanced immune response to rotavirus [26]. This may account for the shortened course of diarrhoea seen in treated patients. The underlying mechanisms of immune stimulation are not well understood, but specific cell wall components or cell layers may act as adjuvant and increase humoral immune responses.

Reduction of diarrhea by probiotics was studied frequently, because diarrhea is the main problem of piglets during the first weeks after weaning with utmost importance for production [54].

The mucosal surface of the intestinal tract represents the largest interface between the body and its environment. An effective local immune is necessary to protect the organism against the invasion of noxious antigens and microbes [54]. No other organ of the body harbours more immune cells than the gut –associated lymphoid tissue (GALT), and a tremendous amount of antibodies is secreted into the intestinal lumen to neutralize and exclude harmful antigens. In numerous studies it has been shown that bacterial colonization influences the function of immune cells belonging to the GALT and even affects the systemic immune system [60].

Immune suppression has been observed after associating germfree rodents with defined bacterial species [69], [50]. In some studies the inductions of immune suppressive cytokines have been implicated in the so-called "by stander suppression" [7]. Moreover, it has been shown that bacterial colonization contributes to the induction and maintenance of immunological tolerance against nutritional antigens [39]. The mechanisms underlying oral tolerance are largely unknown by [54].

Table 2. Incidence of diarrhoea in piglets fed probiotic supplemented feed (Effects compared to control animals) [**54**].

Probiotic	Age	Incidence of diarrhoea	Statistical significance	Literature
B. cereus	8 weeks	Reduced	+	[29]
B. cereus	Day 1-85	Reduced	+	[22]
B. cereus	Day 7-21	Reduced	+	[68]
B. cereus	Day 24-66	No effect	-	[10]
B. cereus	25 kg Live weigh	No effect	-	[27]
B. cereus	2 weeks post weaning	Reduced	+	[23]
E faecium	Day 1-70	Reduced	+	[35]
E. faecium	8 Days before/after weaning	Reduced	+	[51]
P. acidilactici	Day 5-28	Reduced	+	[9]
P. acidilactici *S. cerevisiae*	Day 5-28	Reduced	+	[9]

The numerous studies have reported immune stimulating abilities for different bacterial species. For example, *in vitro* cytokine production of macrophages was stimulated by *Bifidobacteria* [36].*Bifidobacterium longum* as well as several other lactic acid bacteria have been found to increase the total amount of intestinal IgA [57], [65]. *Lactobacillus casei* was reported to have immune adjuvant activity by [43] and *Lactobacillus plantarum* was shown to increase antibody production against*Escherichia coli.* Induction of cytokine profiles by lactobacilli is likely to be strain-dependent [31] and it probably also depends on the host examined, since the autochthonous flora varies between different host species. Most of the animal studies with such probiotic micro organisms have been carried out in rodents with lactic acid bacteria with the goal of designing "functional food" for human consumption. Such studies however, are not necessarily suitable or transferable for the supplementation of animal feed in industrial settings [54]. Studies using swine as model system are few but, seem to be promising.

Probiotic treatment using *Bifidobacterium lactis* HN019 reduced weanling diarrhea associated with rotavirus and *Escherichia coli* infection in a piglet model [52]. Information from studies is also available about the age-dependent development of different immune cells in the intestine of the newborn and adult pigs [62], [55], [56]. Studies on these cells require large amounts of intestinal tissue that can hardly be taken from rodents. The composition of the different immune cells in the GALT is drastically changing during the first the first few weeks of life. For instance, the proliferation rate of B cells in the Peyer's Patches shows a 15-fold increase between days 1 and 42 [56]. Very few observations have been made concerning the influence of bacteria on the development of these immune cells which are the first line of defense against Intestinal infections [54].

A group of authors [54] found a decrease in $CD8^+$ intraepithelial lymphocytes in piglets after treatment of sows and their piglets with *Enterococcus faecium* present in the feed. Neither total IgG or IgA levels in the sera of sow and piglets was affected, nor were the amounts of total IgG or IgA in the milk of the sows influenced by the probiotic treatment. Despite these observations,while the total numbers of coliform bacteria was the same in both probiotic and control herds, there appeared to be at least a 50% reduction in the numbers of pathogenic serovars in piglets from the probiotic group although the rate of isolation of these same serovars in sows was the same for both groups. ELISA-tests to detect specific antibodies against certain pathogenic *Escherichia coli* serovars are still ongoing.

9. OTHER EFFECTS OF PROBIOTICS

Several studies indicate that in pig's intestinal morphology and function of the epithelium may be modified by probiotics [54]. In two trials significantly longer willi were measured in the jejunum of pigs receiving diets supplemented with *Bacillus cereus* [28] and *Bacillus cereus toyoi* or*Saccharomyces boulardii* respectively [17].

The microstructure of the epithelium is of great functional importance for nutrient transport (absorption and secretion) as well as maintenance of transcellular and paracellular barrier functions. This structure inhibits uncontrolled passage of substances and provides a barrier against infection with intestinal bacteria. Carbohydrate structures on the mucosal surface are used for adhesion by pathogenic and non pathogenic bacteria. *In vitro* studies also indicate that some probiotics *Lactobacillus plantarum* 299v and *Lactobacillus rhamnosus GG* have the ability to inhibit adherence of attaching and effacing of pathogenic *Escherichia coli* HT 29 to intestinal epithelial cells by increasing expression of the intestinal mucins MUC2 and MUC3, [32].

Table 3. Some probiotics used as feed additives in European countries [**59**]

The probiotic product	Composition of microorganisms	Utilization
Toyocérine	*Bacillus toyoi*	In all animals
Paciflor	*Bacillus cereus CIP 5832*	In all animals
Adjulact standart	*Enterococcus spp, Lb. lactis, Lb. helveticus, Lb. acidophilus*	Calfs, piglets
Adjulact 1000	*Lb. helveticus, Enterococcus spp*	Calfs, piglets
Adjulact 2000	*Enterococcus spp, Lb. plantarum.*	Calfs,piglets
Yea -sacc	*Saccharomyces cerevisiae*	Ruminants
Lacto-sacc	*Saccharomyces cerevisiae Lb. acidophilus Ec. faecium*	In all animals
Fermacton	*Lactobacillus spp. Ec. faecium* SF68 *Pediococcus spp*	In all animals
Bio-Plus Porc	*Lactobacillus spp. Ec. faecium* SF68 *Pediococcus spp*	Pigs
Lyobacter P1	*Lb. plantarum. Ec. faecium Lb. rhamnosus*	In all animals
Lyobacter SFL	*Ec. faecium* SFL	In all animals
Multigerm	*Lb. plantarum. Ec. faecium Lb. acidophilus*	Pigs
Biosaf SC 47	*Saccharomyces cerevisiae* SC 47	In all animals, especially in ruminants
Bio-Plus 2B	*B. subtilis B. licheniformis*	In all animals
Enteroferm	3 kind of *Lactobacillus, Enterococcus spp. Saccharomyces*	In all animals
Degeferments	*Lb. acidophillus, Lb. lactis*	In all animals
Bacteriolact	*Lb. casei, Str. thermophilus*	Calfs, piglets, lamb

A group of authors [3], [66] concluded that Intestinal mucosa from pigs which were adopted to diets containing *Bacillus cereus* or *Saccharomyces boulardii* had an increased paracellular barrier function and modified nutrient transport kinetics for glucose and amino acids. For *Lactobacillus plantarum* 299v was shown, that pretreated rats were protected against increase in intestinal permeability induced by*Escherichia coli* [33].

10. EXPERIMENTS IN EXTENSIVE FARM CONDITIONS

10.1. Material and Methods

Two animal trials were carried out at the same private farm of pigs. Twenty four piglets (White x Duroc) of four litters were transferred after weaning (35 days) to flat decks and randomly allocated to 4 groups with 6 animals (3 male and 3 female). The basal diet (see Table 4 and 5) was also supplemented with 1000mg, 1500mg and 2000mg/kg of the probiotic preparation (three experiment groups) or without supplementation (control group). The diets were offered ad-libidum and animals had free access to water. The probiotic preparation included the following strains: *Lactobacillus plantarum*ATCC 4336 ($5x10^9$ CFU/kg), *Lactobacillus fermentum* DSM 20016 ($5x10^9$ CFU/kg) and *Enterococcus faecium* ATCC 19434 ($5x10^{10}$ CFU/kg) (AKRON s.r.l-Milano). During the eight weeks experimental period in the first experiment and six weeks experimental period in the second experiment, body weight (BW), daily weight gain (DWG) and feed conversion ratio (FCR), kg feed/kg body weight gain were measured weekly. Data are presented as arithmetic means with standard deviations (Mean ± SD). One-way analysis of variance and Student's t-test ($P< 0.05$) were performed to test the differences between levels of the probiotic in the diet.

10.2. Results and Discussions

In last ten years, most of the experiments were performed with piglets. According to the literature review, in many trials showed positive effects of probiotics on weaned piglets and also there were no significant effects of growing and finishing pigs. In the first trial the body weight gain was improved with graded levels (1000 and 1500 mg/kg feed) of the probiotic preparation respectively 15% to

11%, compare to control group. In the fourth and eighth weeks of this trial, a significant difference was documented. The body weight gain, on the second experiment was improved with graded levels (1000-1500 mg/kg feed) of the probiotic preparation from 3% to 6%, compare to control group, without significance. The FCR (kg feed/kg weight gain) in the first trial was improved with graded levels by up to 13.3%, 11.3% and 0.4% compare to control group and in the second trial respectively 12.5%, 10.4% and 8.5% compare to control group. The tendency for increasing of probiotic dose has not positive effects on performance parameters. Because of the low dose-response between 1000 and 1500 mg/kg feed, the level of 1000 mg/kg feed seems to be the optimal dose [64].

According to [20] on the experiments with weaned pigs and growing-finishing swine, used 1g/kg*Lactobacillus acidophilus,* which contains $4x10^6$ viable cells per gram. Supplementation of the diet with 1g/kg *Lactobacillus acidophilus* on weaned pigs did not improve daily gain, feed intake or feed efficiency. Daily weight gain and feed intake of pigs, treated with 500 mg/kg *Lactobacillus acidophilus*showed non significant trends.

Reduction of diarrhoea by probiotics and vitality of piglets is one of the second topics in this study, because diarrhoea is the main problem for weaned piglets, especially during the first week after weaning. After two weeks of probiotic supplementation, we showed a reduction of diarrhoea on three treated groups. Reduction of diarrhoea by probiotic supplementation was study frequently by many scientist groups. Some of the trials showed significant effects, but the others have collected not significant data. A group of authors [29], [22], [68], [23] have used the same probiotic *Bacillus cereus* in different age of piglets, respectively 8 weeks piglets, 1-85 day after birth, 7-21 day after birth and 2 weeks post weaning. They showed statistical significance of diarrhoea reduction. [10] showed non significant effects, while they used *Bacillus cereus* in pigs 24-66 days of life.

Table 4. Diet composition and calculated nutrient concentration on the first experiment.

Diet composition (g/kg feed)		Nutrient concentration (g/kg feed)	
Maize	620	ME (MJ/kg)	12,33
Soya bean meal	280	Crude protein	196.4
Sunflower meal	50	Crude fat	28,70
Fish meal	10	Crude fibre	42,90
Limestone	15	Calcium	10,77
Monocalcium phosphate	15	Phosphorus	6,50
Vitamin -mineral premix[a]	5	Lysine	11,30
L-Lysine	5	Methionine+Cystine	6,70

[a] Contents in 1 kg: 1,200,000 IE vit. A, 120,000 IE vit. D_3, 4000 mg vit. E, 200 mg vit. B_1, 600 mg Vit. B_2, 2500 mg Niacin, 400 mg Vit. B_6, 4500 µg Vit. B_{12}, 20,000 µg Biotin, 1800 mg Pantothenic acid, 160 g Na, 50 g Mg,10,000 mg Zn, 7500 mg Fe, 7500 mg Mn, 150 mg J, 70 mg Co and 40 mg Se.

Table 5. Diet composition and calculated nutrient concentration on the second experiment.

Diet composition (g/kg feed)		Nutrient concentration (g/kg feed)	
Maize	630	ME (MJ/kg)	12,90
Soya bean meal	320	Crude protein	197,1
Fish meal	10	Crude fat	28,08
Limestone	10	Crude fibre	35,94
Monocalcium phosphate	15	Calcium	8,60
Vitamin-mineral premix	10	Phosphorus	6,72
L-Lysine	5	Lysine	10,65

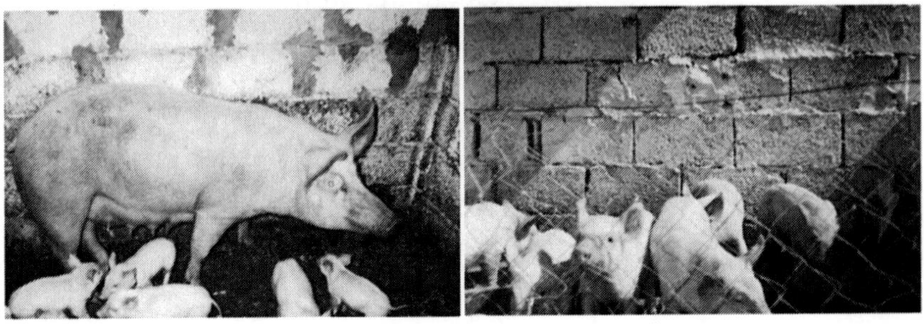

Figure 1. Piglets in the first and second experiments, in extensive farm condition.

11. EXPERIMENT IN INTENSIVE FARM CONDITION

11.1. Material and Methods

The animal trials were carried out at the experimental station of the Institute of Animal Nutrition of the Free University of Berlin, Germany. Thirty two piglets (White x Duroc) of three litters were transferred after weaning (28 days) to flat-decks and randomly allocated to 4 groups with 8 animals (4 male and 4 female). The basal diet was either supplemented with 1000, 1500 and 2000 mg/kg of the probiotic preparation or without supplementation (control).

The diets were offered ad libitum and animals had free access to water. The probiotic preparation included the following strains: *Lactobacillus plantarum* ATCC 14917 $1x10^{11}$ CFU/kg, *Lactobacillus fermentum* DSM 20016 $1x10^{11}$ CFU/kg and *Enterococcus faecium* ATCC 19434 $1x10^{11}$ CFU/kg. During the six weeks period body weight (BW), daily weight gain (DWG) and feed conversion ratio (FCR), kg feed/kg body weight gain were measured weekly. Three piglets from each trial group were euthanized one week after probiotic administration by intracardial injection of T61 (Fa. Hoechst) after sedation with Stresnil*. Immediately after death, the abdomen was opened and ligatures were applied to collect digesta samples for pH measurement in defined segments of the duodenum, jejunum, ileum, caecum and colon. This operation was finished between 12-14 hours after death.

For determination of intestinal bacteria, the "Selective Media" method was used (CATC-agar (Citrat Acid Tween Carbonate - agar base) for *Enterococci spp*, MRS-agar (*Lactobacillus* agar acc to Man Rogosa and Sharp) for *Lactobacilli spp* and Mac Conkey for *Enterobacteria spp*). The colony of *aerobe and anaerobe* micro organisms by visual numbering were measured on agar plate.

The apparent nutrient digestibility was determined by the indicator method during the last week of the experiment using chromium (III) oxide (0.5%).

$$\text{Coeficient of digestibility} = 100 - \left(\frac{\%\text{ e indicator in feed x }\%\text{ e nutrient in faeces}}{\%\text{ e indicator in faeces x }\%\text{ e nutrient in feed}} \text{x } 100 \right)$$

Data are presented as arithmetic means with standard deviations (Mean ± SD). One-way analysis of variance and Student's t-test ($P< 0.05$) were performed to test the differences between levels of the probiotic in the diet.

Table 6. Effects of probiotic preparation on performance parameters in the first experiment (Mean ± SD).

Parameters		Probiotic Dose (mg/kg feed)			
		0	1000	1500	2000
Production	n[1]				
Initial BW, kg	6	5.3 ± 0.65	5.4 ± 0.77	5.6 ± 0.37	5.1 ± 0.17
Fourth weeks[4]		12.59 ± 2.63	14.20 ± 1.62[a]	13.93 ± 0.82	10.97 ± 0.93[b]
Eighth weeks		19.89 ± 2.05	23.00 ± 2.73[a]	22.26 ± 2.42	18.84 ± 1.43[b]
DWG, g [2]		260.7 ± 33.8	314.3 ± 62.9[a]	297.6 ± 71.6	245.4 ± 46.5[b]
FCR [3]		3.01 ± 0.68	2.61 ± 0.25	2.67 ± 0.32	2.94 ± 0.42

[1]Number of animals/every group

[2] DWG for whole experimental period.

[3] FCR for whole experimental period.

[4] Significant differences, indicated with different superscripts.

Table 7. Effects of probiotic preparation on performance parameters in the second experiment (Mean ± SD).

Parameters		Probiotic Dose (mg/kg feed)			
		0	1000	1500	2000
Production	n[1]				
Initial BW, kg	6	4.8 ± 0.65	5.1 ± 0.77	5.0 ± 0.37	4.9 ± 0.17
Sixth weeks		16.37 ± 3.76	17.37 ± 4.06	16.98 ± 3.98	16.25 ± 3.45
DWG, g[2]		275.6 ± 46.7	292.3 ± 57.3	285.4 ± 51.8	270.4 ± 43.7
FCR [3]		3.20 ± 0.76	2.80 ± 0.48	2.87 ± 0.57	2.93 ± 0.68

Table 8. Diet composition and calculated nutrient concentration.

Diet composition (g/kg feed)		Nutrient concentration (g/kg feed)	
Maize	620	ME (MJ/kg)	12.82
Soybean meal	275	Crude protein	197.8
Soya oil	50	Crude fat	34.3
Fish meal	30	Crude fibre	31.4
Limestone	10	Calcium	9.10
Monocalcium phosphate	15	Posphorus	7.68
Vitamin -mineral premix[a]	12	Lysine	11.77
L-Lysine	10	Methionine+Cystine	7.64
Methionine+cystine	10	Threonine	8.04
Threonine	10	Tryptophane	2.37
Tryptophane	3		

[a] Contents in 1 kg: 1,200,000 IE vit. A, 120,000 IE vit. D_3, 4000 mg vit. E, 200 mg vit. B_1, 600 mg Vit. B_2, 2500 mg Niacin, 400 mg Vit. B_6, 4500 µg Vit. B_{12}, 20,000 µg Biotin, 1800 mg Pantothenic acid, 160 g Na, 50 g Mg,10,000 mg Zn, 7500 mg Fe, 7500 mg Mn, 150 mg J, 70 mg Co and 40 mg Se.

12. THE METHODOLOGY FOR DETERMINATION OF MICROBIOLOGICAL CHARGE OF FAECES

Microbiological analyzes of faeces were performed in two periods:

- Week 1-3
- Week 5-7

In the first period, such analysis aim to consistently follow microbiological changes due to the "probiotics" effect.

In the second period, such analysis aim to compare the microbiological changes in the beginning and in the end of the experiment, as well as to judge on the duration of the "probiotics" effect after its termination.

Microbiological analyses were carried out of as follows:

3-4 hours after the feed, fresh faeces was collected in plastic boxes. Faeces of all boxes were gathered and placed in a separate box. 1 g of faeces was taken for each box, in three parallel tests A, A_1, A_2.

9 ml Ringer solution was added to it, and the following dilutions were prepared:

10^{-1}-10^{-9}, MRS for identification of *Lactobacillus spp*

10^{-4}-10^{-8}, CATC for identification of *Enterococcus spp*

10^{-3}-10^{-8}, McK for identification of *Enterobacteriaceae*

Its cultivation in Agar plates and incubation at a temperature of 37^0C was conducted within 24 hours.

13. THE PHYSIOLOGICAL AND MICROBIOLOGICAL PARAMETERS OF INTESTINAL MUCOSA AND DIGESTA

A week after administration of probiotics, a total of 12 piglets were slaughtered, 3 piglets for every group.

The slaughtering of pigs a week after administration of probiotics aimed at:

- monitoring of the changes occurring in the pH digesta in the intestines.
- monitoring of all microbiological changes in digesta and mucosa, reflecting *Lactobacillus spp,Enterococcus spp* and *Escherichia coli* microbiological load as well as the total number of anaerobic bacteria in the jejunum, ileum, caecum and colon.

The preparation of samples for microbiological analysis was carried out as follows:

A 2x10cm area from all parts of intestine and colon is taken. Then, it is washed away with 0.9% NaCl solution, is measured its length, is thorn with a fine scalpel, is weighed and finally is placed in plastic tubes. Since jejunum is relatively long, it is divided into three parts for more convenience: jejunum 1, jejunum 2 and jejunum 3.

Measuring and weighing was done for the following parts:

Duodenum
Jejunum 1
Jejunum 2
Jejunum 3
Ileum
Caecum
Colon

Microbiological load was estimated at:

Middle of jejunum, ileum, caecum, beginning of colon

14. THE DETERMINATION OF ANAEROBIC BACTERIA (*LACTOBACILLUS SPP)*. METHOD OF SAMPLES IN ICE

15 ml digesta is taken, is squeezed, and after is being cast into sterile plastic tubes and it is placed in ice.

0.5 g of this digesta is taken, 500 ml Ringer solution is added, and then is placed on ice.

Dilutions are prepared by mixing what is taken from both beakers up to 100l.

20l is taken by pipette and is dripped in Agar plates prepared based on the following dilutions:

MRS: 10^{-6} to 10^{-10} Columbia - Blut: 10^{-6} to 10^{-10}

15. METHODS OF SAMPLES IN ICE

Parts of the intestines are cut and placed in 50ml tubes together Ringer solution. Later solution is shaken and changed until no more digesta remains. The prepared solution is put into a bottle and placed in ice. Intestine is placed on a plate, mucosa is thorn and mixed. 0.5 g mucosa is taken; 500l Ringer solution is added and placed on ice.

Dilutions are prepares as in the first case and are placed on ice.

201 is taken by pipette and transferred to Agar plates prepared according to the following dilutions:

MRS: 10^{-5} to 10^{-9} Columbia –Blut: 10^{-5} to 10^{-10}

16. THE DETERMINATION OF AEROBIC BACTERIA (*ENTEROBACTERIACEAE* AND *ENTEROCOCCUS SPP*)

Digesta dilutions are prepared as above. 201 solution is taken and transferred to Agar plates prepared according to the following dilutions:

Mac Conkey: 10^{-6} to 10^{-10} CATC : 10^{-3} to 10^{-7}

Mucosa dilutions are prepared. 201 solution is taken and transferred to Agar plates prepared according to the following dilutions:

Mac Conkey : 10^{-3} to 10^{-7} CATC : 10^{-2} to 10^{-6}

Microbiological load was estimated: Middle of jejunum, ileum, caecum, beginning of colon

17. DATA ABOUT PROBIOTIC "*SEBERINI SUINI*"

17.1. Microbiological Composition Of Probiotic

*Lactobacillus plantarum*ATCC 14917 (LMG – S 16691) cfu 1x 10^{11}

*Lactobacillus fermentum*DSM 20016 (LMG- S 16517) cfu 1x 10^{11}

Figure 2. Institute of Animal Nutrition, Free University, Berlin

Figure 3. The animal trial at the experimental station of the Institute of Animal Nutrition.

*Enterococcus faecium*ATCC 19434 (LMG- S 16690) cfu 1x 10^{11}

Composition of the probiotic "Seb Suini"

Lactobacillus plantarum	25 %
Enterococcus faecium	10 %
Lactobacillus fermentum	15 %
Micronized soya extraction meal	50 %

Table 9. Chemical composition of the probiotic "Seb Suini" used in the experiment.

Chemical compositon %	Amino acids g/kg
Dry matter 88	Lysine 17
Crude protein 35	Leucine 17
Crude fat 1	Threonine 11
Crude fibre 5	Arginine 10
Crude ash 28	Tryptophan 3
	Izoleucine 11
	Hystidine 6
	Glycine 9
	Cystine 2
	Valine 13

18. PHYSICAL -CHEMICAL CHARACTERISTICS OF THE PROBIOTIC

Smell	tipical, not bad
Apparent densities after shaking	0,45 kg/liter.
Point of degradability	> 250°C
Density	450 gr/liter
Water solubility	non digestible, hydrodispersible.
Granulometry	90% e grimcave kalojnë sitën 200 micron.
Value of pH	6,5 (10 gr on 100 ml in temperature 20°C)
Microbiological characteristics	
Total not lactic flora	maximum 5 x 10^3 UFC/gr
Enterobacteriaceae	absent
Coliformes	absent
Enterococcus	maximum 5 x 10^2 UFC/gr
Yeasts and moulds	maximum 1 x 10^2 UFC/gr

According to the analyzes made in the Institute of Soil Chemistry, "Universitá Cattolica del Sacro Cuore"- Piacenza, results heavy metal contain

Pb	<0,6	ppm
Cd	94	ppm
Ni	11	ppm
Cr	15	ppm
As	1, 18	ppm
Hg	112	ppm

It does not contain Alfa-toxine B_1, B_2, G_1, G_2, Zearalenone, Ocratoxine, Fumosine B_1,

Deossinivalenolo 122,0g / Kg tq

19. RESULTS AND DISCUSSIONS

The body weight gain was improved with graded levels of the probiotic preparation from 4.9 up to 31.7%. Caused by the high coefficient of variation the differences were not significant. The FCR (kg feed/kg weight gain) was improved with graded levels by 0.6 up to 7.3%. The differences were not significant. Because of the low dose-response between 1500 and 2000 mg/kg feed, the level of 1500 mg/kg feed seems to be the optimal dose.

The same results showed [30] on the experiments with weaned piglets, used LFP- *Lactobacillus-Fermentation-Product.* This probiotic contents *Lactobacillus bulgaricus, Lactobacillus casei, Streptococcus thermophilus*, produced in Quebec, Canada. The basal diet was supplemented with 100 mg LFP/kg feed.

The feed intake and the daily weight gain (DWG) were increased respectivly 11.8% and 10.4%, compared with the control group. The feed conversion ratio (FCR) was in the same level.

Two authors [19] used the same probiotic LFP (*Lactobacillus-fermentation-product*) on the weaned piglets. Pigs fed a diet with 0.36 ml/kg LFP required nearly 10% less feed per unit of weight gain than the control group. Also the incidence of scouring decreased ($P< 0.05$) in pigs fed with different levels of LFP. Overall improvement occurred up through the addition of 0.36 ml/kg LFP with no additional benefit from greater amounts. Another group of authors [44] showed the effects of microbial feed additives on performance of starter and growing-finishing pigs. One of the experimental group with weaned piglets was fed with 750 mg *Lactobacillus acidophilus*/kg feed. The second experimental group was supplemented with 1250 mg *Streptococcus faecium*/kg feed.

Table 10. Effects of probiotic preparation on performance parameters (Mean ± SD).

Parameters		Probiotic Dose (mg/kg feed)			
		Control	1000	1500	2000
Production	n[1]				
-Initial BW, kg	8	5.6 ± 1.11	5.5 ± 1.07	5.6 ± 1.17	5.6 ± 1.02
-BW 6th week [2]	5	19.5 ± 5.10	19.8 ± 5.83	23.1 ± 3.17	22.3 ± 7.01
Feed intake, kg		24.5 ± 7.49	25.4 ± 6.44	29.79 ± 5.42	30.4 ± 7.47
DWG, g [3]		325 ± 153	341 ± 128	427 ± 71	436 ± 123
FCR [4]		1.79 ± 0.48	1.78 ± 0.31	1.65 ± 0.05	1.66 ± 0.15

[1] Number of animals, (8 piglets/ every group, at the beginning of the experiment)

[2] Number of animals, (5 piglets/every group, one week after probiotic supplementation). n = 4 at treatment 1500 mg/kg in 6th week. [3]DWG for whole experimental period. [4]FCR for whole experimental period.

The addition of *Lactobacillus acidophilus* to the feed of young pigs improved average daily weight gain by 9.7 % and the feed conversion ratio by 21.4%, whereas the addition of *Streptococcus faecium*decreased average daily weight gain. The addition of acid lactic improved feed conversion, suggesting that lactic acid as a metabolite produced during fermentation might be the reason for the improvement in performance. The probiotics had no effect on growing-finishing pigs.

In a trial with 90 untreated and 90 treated *(Bacillus cereus*-preparation) weaned piglets, the probiotic treated animals gained 7% more live weight during 6 weeks after weaning with a reduced feed conversion ratio of 2.4%. However, both results were not significant. This points towards a high variation in the response of the individual animals to this type of feed additives [23].

Table 11. Effects of probiotic preparation on apparent nutrient digestibility and digesta pH of defined intestinal segments (Mean ± SD).

	N[1]	Control	1000	1500	2000
Digestibility [2] (in %)	5				
Dry matter		76.4 ± 6.90	73.2 ± 10.39	67.2 ± 2.22	75.7 ± 9.52
Crude fat		75.1 ± 5.48	71.2 ± 2.60	69.0 ± 9.11	70.0 ± 3.77
Crude fibre		51.1 ± 7.82	54.5 ± 7.48	52.3 ± 5.79	56.4 ± 2.31
Digesta pH	3				
Duodenum		5.54 ± 0.96	5.74 ± 0.68	5.87 ± 0.83	6.51 ± 0.77
Jejunum		6.24 ± 0.38	6.17 ± 0.66	6.29 ± 0.51	6.56 ± 0.85
Ileum[3]		7.05 ± 0.43[a]	6.43 ± 0.77[b]	6.41 ± 0.16[b]	5.25 ± 0.12[c]
Caecum		5.62 ± 0.13	5.65 ± 0.20	5.79 ± 0.39	5.55 ± 0.09
Colon		5.87 ± 0.27	6.19 ± 0.38	6.27 ± 0.37	6.18 ± 0.43

[1] Number of animals.

[2] Crude nutrients were determined by Weende scheme.

[3] Significant differences, indicated with different superscripts.

Feeding probiotic preparation slightly increased the crude fiber digestibility compared to the control group in the range of 3.4%, 1.2% and 5.4% at supplementations with 1000, 1500 and 2000 mg/kg feed, respectively. With graded levels of the probiotic preparation pH of the chyme of ileum and caecum was slightly decreased, in contrast the pH of duodenum and jejunum was slightly increased [63]. The low effect of pH was agreement with digestibility results. The pH results in the duodenum and jejunum is in contrast to former results reported by [35]. This is possibly caused by the combination of different strains used in this study.

Two authors [19] supplemented the diets of growing pigs with LFP preparation (*Lactobacillus Fermentation Produced*) and observed that a supplementation of 0.72 mg LFP/kg feed increased the crude fibber digestibility with 14.2% compared to the control group ($P< 0.05$).

These authors assumed that the rate of passage of feed through the digestive tract was decreased by feeding LFP, which allowed more time for digestion of crude fiber. Also the urinary nitrogen excretion was greater than faecal excretion but both combined were less then intake, thus resulting in a positive nitrogen balance. In total, the digestibility of dry matter was decreased 0.4% and the digestibility of crude protein did not change, compared to the control. Another author [58] showed the influence of*Lactobacillus acidophilus* in broïler chicks on growth, feed conversion and crude fat digestibility. The addition of *Lactobacillus acidophilus* in broïler chicks diet decreased the digestibility of crude fat.

The effect of probiotic preparation on the microbial composition of faeces was examined early, one week after supplementation, because the first week after weaning is critical period for tends to shift the balance of the gut microflora away from beneficial bacteria towards pathogenic bacteria. One week after weaning piglets fed with the probiotic preparation showed increased the concentration of*Lactobacilli* spp. and *Enterococci spp.* compared to the control treatment. Feeding 2000 mg probiotic preparation/kg feed induced a reduction of *Escherichia coli.* At the end of the experiment piglets fed with 1500 and 2000 mg probiotic preparation/kg feed had reduced *Escherichia coli* compared to the control. These results indicate that the probiotic preparation may be less suppressive to the *Escherichia coli.* [40] observed the similar microbial changes in the faeces of weaned piglets, fed with the same combined probiotic preparation.

Table 12. The effect of probiotic preparation on the microbial composition of faeces (CFU*10^6/g wet weight) (Mean ± SD).

			Probiotic Dose (mg/kg feed)		
		Control	1000	1500	2000
1st week of trial	*Lactobacilli* spp.	95	120	150	170
	Enterococci spp.	0.01	0.94	1.12	1.23
	Escherichia coli.	10	10	32	2
6th weeks of trial	*Lactobacilli* spp.	683 ± 584	223 ± 191	345 ± 403	767 ± 306
	Enterococci spp.	0.018 ± 0.031	0.1 ± 0.131	0.011 ± 0.01	0.028 ± 0.02
	Escherichia coli.	2.35 ± 3.60	15 ± 21.8	0.05 ± 0	0.083 ± 0.057

* Four faeces samples/every group were collected/every week, during the experimental period.

The effects of the probiotic preparation on the microbial composition of the chyme showed no dose–depended effects. However there was a tendency for increasing of the concentration of *Lactobacilli*spp. and *Enterococci* spp. in the colon compared to the control.

Table 13. The effect of probiotic preparation on the microbial composition of digesta, one week after probiotic supplementation. (log CFU/g wet weight), (Mean ± SD; n = 3).

		Control	Probiotic Dose (mg/kg feed)		
			1000	1500	2000
Jejunum	*Anaerobe bacteria.*	13.92 ±14.15	12.22 ± 12.45	8.75 ± 8.60	12.98 ± 13.07
	Lactobacilli spp.	10.24 ± 10.44	12.58 ± 12.78	8.36 ± 8.38	11.60 ± 11.55
	Enterococci spp.	7.02 ± 6.98	8.03 ± 8.22	7.00 ± 7.19	7.01 ± 6.97
	Escherichia coli.	7.57 ± 7.74	8.60 ± 8.72	6.00 ± 0.00	7.90 ± 8.02
Ileum	*Anaerobe bacteria.*	13.17 ± 13.36	13.21 ± 13.20	13.21 ± 13.20	12.60 ± 12.72
	Lactobacilli spp.	12.87 ± 13.11	12.69 ± 12.73	12.72 ± 12.95	13.68 ± 13.89
	Enterococci spp.	6.00 ± 0.00	8.82 ± 9.06	7.33 ± 7.55	7.02 ± 7.22
	Escherichia coli.	8.17 ± 8.17	11.00 ± 11.23	12.01 ± 12.25	12.05 ± 12.23
Caecum	*Anaerobe bacteria.*	13.90 ± 13.85	12.69 ± 12.84	13.75 ± 13.87	13.98 ±14.12
	Lactobacilli spp.	13.28 ± 13.48	12.60 ± 12.84	13.43 ± 13.65	13.83 ± 14.05
	Enterococci spp.	6.86 ± 7.04	10.00 ± 10.23	7.80 ± 8.03	6.84 ± 6.70
	Escherichia coli.	12.69 ± 12.93	10.00 ± 10.23	10.82 ± 11.06	10.86 ± 11.04
Colon	*Anaerobe bacteria.*	14.72 ± 14.92	13.04 ± 13.06	13.95 ± 14.18	13.93 ± 14.15
	Lactobacilli spp.	12.55 ± 12.49	13.01 ± 13.23	13.84 ± 14.08	13.92 ± 14.10
	Enterococci spp.	8.82 ± 9.06	9.00 ± 9.23	12.01 ± 12.25	9.12 ± 9.36
	Escherichia coli.	13.44 ± 13.68	11.30 ± 11.53	12.69 ± 12.93	12.39 ± 12.59

A group of authors [1] supplemented the pig diets with a combination of *Lactobacillus fermentum* 14 and *Streptococcus salivarius* 312 for 4 days and observed a significant reduction in the *Escherichia coli*count in both the stomach and duodenum. A significant reduction of *Escherichia coli* number in the stomach was also found, when *Lactobacillus fermentum* was supplemented separate. In cases of diarrhoea caused by *Escherichia coli* the treatment as described here was not effective because the count of *Escherichia coli* in the duodenum of culture-fed pigs was still greater than 10^6/g. However, if the antibacterial effect of strain 14 could be increased some effect on scouring due to *Escherichia coli*should follow. This might be accomplished by the feeding of large numbers of organisms or by the administration in a concentrated form of the inhibitory factors produced by *Lactobacillus fermentum*strain 14. [15] showed that the application of 10^8 colony forming units (CFU) of a *Bacillus cereus*preparation/kg feed to piglets reduced counts for *Lactobacilli spp. Bifidobacteria*, *Eubacteria* and*Escherichia coli* in the duodenum and jejunum, but increased respective CFU in the ileum, caecum and colon.

Two authors [35] showed a significant reduction of *Escherichia coli* CFU in the small intestine of piglets was also noted when an *Enterococcus faecium* preparation was applied. However, at the same time *Lactobacilli spp.* and *Enterococci spp.* counts increased as a trend and statistically significant, respectively [24].

The results of studies on the ability of probiotic bacteria to reduce the colonization of pathogenic bacteria are ambiguous. Challenge studies with piglets and *Escherichia coli* O141:K85 showed no influence on clinical symptoms, mortality or excretion of hemolytic *Escherichia coli* [8]. A group of authors [24] showed that the colonization with mucosa associated *Enterobacteria spp.* was reduced when a probiotic *Bacillus cereus* preparation was supplemented.

The probiotic had no influence on the occurrence of pathogenic *Escherichia coli* as measured with a PCR assay [16]. These results point to the fact that hygienic conditions in scientific institutes may sometimes be too favorable to investigate effects of pathogenic bacteria without challenge trials [54].

These and the other studies imply that probiotics are able to reduce/enhance specific bacterial groups, but the reduction of total bacterial cell numbers as recorded for antibiotics is probably not a probiotic mode of action. In order to understand the casual relationships which lead to the observed improvements in weight gain and feed conversion or general health of animals, possible interactions between bacteria in the intestine and host animal must be studied. Of special significance are interactions between the metabolism of the host and metabolic activity of intestinal bacterial populations [54].

20. CONCLUSIONS

The supplementation of the combined probiotic preparation induced slightly the performance data. In extensive farm condition, a significant difference of daily weight gain (DWG) was documented four weeks after probiotic supplementation. A positive effect of the probiotic on feed conversion ratio (FCR), kg feed/kg weight gain and vitality was observed, also. We recommend the level of 1000mg/kg feed combined probiotic as the optimal dose.

Combined probiotic preparation induced slightly the performance data in intensive farm condition, also. However the differences were not significant. Feeding probiotic preparation slightly increased the crude fibre digestibility in all treated groups. With graded levels of the probiotic preparation pH of the chyme of ileum and caecum was slightly decreased, in contrast the pH of duodenum and jejunum was slightly increased. The probiotic preparation showed increased the concentration of *Lactobacilli spp.*and *Enterococci spp.* compared to the control. The results indicate that the probiotic preparation may be less suppressive to the *Escherichia coli.* The effects of the probiotic preparation on the microbial composition of the chyme showed no dose–depended effects. However there was a tendency for increasing of the concentration of *Lactobacilli spp.* and *Enterococci spp.* in the colon compared to the control. Possibly this was due to the combined probiotic preparation. At the end, we recommend the level of 1500 mg/kg feed combined probiotic as the optimal dose.

* Approved by competent authority according to Council Directive 86/609/EEC of 24 November1986 on the approximation of laws, regulations and administrative provisions of the Member States, regarding the protection of animals used for experimental and other scientific purposes.

ACKNOWLEDGEMENT

The authors are grateful to Dr. K. Schäffer and all technicians stuff for technical assistance. Research stay of Dr. E. Delia in Institut für Tierernährung, Freie Universität, Berlin, Germany was financial supported by Deutsche Gesellschaft für Technische Zusammenarbeit (GTZ) and Tempus Phare Project "Animal Science Albania" AC_JEP-14123-1999.

REFERENCES

1. P. A. Barrow, B. E. Brooker, R. Fuller, M. J. Newport, 1980The attachment of bacteria to the gastric epithelium of the pigs and its importance in the microecology of the intestineJ. Appl. Bacteriol. 48147154

2. M. F. Bernet-Camard, M. H. Coconnier, S. Haudault, A. L. Servin, 1996Differentiation-associated antimicrobial functions in human colon adenocarcinoma cell lines. Exp. Cell. Res, 2268089

3. G. Breves, C. Walter, M. Burmeister, B. Shröder, 2000In vitro studies on the effects of Saccharomyces boulardii and Bacillus cereus var. toyoi on nutrient transport in pig jejunumJ. Anim. Physiol and Anim Nutrition. 84920

4. P. H. Brooks, J. D. Beal, V. Dmeckova, S. Niven, 2003Probiotics for pigs and beyond. In: Van Vooren and B. Rochet. Role of probiotics in animal nutrition and their link to the demands of Europian consumers, ID-Lelystad, 4959

5. I. Castagliuolo, M. F. Riegler, L. Valenick, J. T. La Mont, C. Pothoulakis, 1999Saccharomyces boulardii protease inhibits the effects of Clostridium difficile toxins A and B in human colonic mucosaInfect. Immun. 67302307

6. Collins M.D, Gibson G.R1999Probiotics, prebiotics and synbiotics: approaches for modulating the microbial ecology of the gut. Am. J. Clin Nutr. 69 (Suppl):1052S-1057S.

7. A. Dahlman-Hoglund, L. A. Hanson, S. Ahlstedt, 1997Induction of oral tolerance with effects on numbers of IgE-carrying mast cells and on bystander suppression in young ratsClinical Experimental Immunology 108128137

8. F. De Cupere, P. Deprez, D. Demeulenaere, E. Muylle, 1992Evaluation of the effect of 3 probiotics on experimental Escherichia coli enterotoxaemia in weaned piglets.J. Vet. Med. Bulletin. 39277284

9. L. Durst, M. Feldner, B. Gedek, B. Eckel, 1998Bakterien als Probiotikum in der Sauenfütterung und der Ferkelaufzucht. Kraftfutter 9356364

10. U. Eidelsburger, M. Kirchgessner, F. X. Roth, 1992Zum Einfluss von Fumarsäure, Salzsäure, Natriumformiat, Tylosin und Toyocerin auf tagliche Zunahmen, Futteraufnahme, Futterverwertung und Verdaulichkeit: 11. Mitt. Journal of Animal Physiologie and Animal Nutrition. 688292

11. M. Freitag, H. Hensche-U, H. Schulte-Sienbeck, B. Reichelt, 1998Kritische Betrachtung des Einsatzes von Leistungsförderern in der Tierernährung.Forschungsberichte der Universität Paderborn, Nr. 8.

12. R. Fuller, 1989Probiotics in man and animals.J. Appl. Bacteriol. 66365378

13. R. Fuller, 1999Probiotics for farm animalsProbiotics: A critical Review.

14. Gaskins H.R1996Immunological aspects of host microbiota interactions at the intestinal epithelium. Gastrointestinal microbiology, 2537587

15. K. Gedek, M. Kirchgessner, S. Wiehler, A. Bott, U. Eidelsburger, F. X. Roth, 1993Zur Nutritiven Wirksamkeit von Bacillus cereus als probiotikum in der ferkelaufzuchtArchiv Anim. Nut. 44215226

16. Goebel S, Vahjen W, Jadamus A, Simon O (2000) PCR assay for detection of porcine pathogenic Escherichia coli virulence factors in the gastrointestinal tract of piglets fed a spore forming probiotic. Proc. 9th Society for Nutrition and Physiology. 64.

17. B. Goerke, 2000Untersuchungen zur Schleimhautmorphologie im Dünn-und Dickdarm nach oraler Applikation von Saccharomyces boulardii und Bacillus cereus var.toyoi beim Schwein. Doctor Thesis, Tierärtzliche Hochschule Hannover, Germany.

18. B. R. Goldin, S. L. Gorbach, M. Saxelin, S. Barakat, L. Gualtieri, S. Salminen, 1992Survival of Lactobacillus species (strain GG) in human gastrointestinal tract.Dig. Dis. Sci. 37121128

19. Hale O.M, Newton K.I1979Effetcs of a nonviable Lactobacillus species fermentation product on performance of pigs. J. Anim. Sci. ;484770775

20. Harper A.F, Kornegay E.T, Brayant K.L, Thoman H.R1983Efficacy of virginiamycin and a commercially-available Lactobacillus probiotic in swine dietsAnim Feed Sci Technol, 86976

21. R. Havernaar, Brink. B. Ten, in't. Huis, J. H. J. Veld, 1992Selection of strains for probiotic use. In: Probiotics. The scientific basis. R.Fuller (Ed.). Chapman& Hall, London, 209224

22. Ch. Iben, J. Leibetseder, 1989Untersuchung der leistungsfördernden Wirkung von Toyocerin in der Ferkelaufzucht. Wien. Tierärztliche Monatsschrift 76363366

23. A. Jadamus, 2001Untersuchungen zur Wirksamkeit und Wirkungsweise des sporenbildenden Bacillus cereus var. toyoi im Verdauungstrakt von Broilern und Ferkel.Degree Dissertation, Free University, Berlin

24. A. Jadamus, W. Vahjen, O. Simon, 2000Influence of the probiotic bacterial strain, Bacillus cereus var. toyoi on the development of selected microbial groups adhering to intestinal mucosal tissues of piglets.J. Anim. Feed Sci.;9347362

25. A. Jadamus, W. Vahjen, O. Simon, 2001Growth behaviour of a spore forming probiotic strain in the gastrointestinal tract of broiler chicken and piglets.Archiv of Anim. Nut. 54117

26. M. Kaila, E. Isolauri, E. Soppi, E. Virtanen, S. Laine, H. Arvilommi, 1992Enhancement of the circulating antibody secreting cell response in human diarrhea by a human Lactobacillus strain.Pediatr. Res. 32141144

27. M. Kirchgessner, R. X. Roth, U. Eidelsburger, B. Gedek, 1993Zur nutritiven Wirksamkeit von Bacillus cereus als probiotikum in der Ferkelaufzuchtmittelung Einfluss auf Wachstumsparameter und gastrointestinales Milieu. Archives of Animal Nutrition44111121

28. U. Klein, H. L. Schmidts, 1997Zum Einfluss des Bioregulators Paciflor auf die Morphologie der Dünndarmmukosa beim Schwein. Proceedings of the Society for Nutrition and Physiology 6:41.

29. S. C. Kyriakis, V. K. Tsiloyiannis, J. Vlemmas, K. Sarris, A. C. Tsinas, C. Alexopoulos, L. Jansegers, 1999The effect of probiotic LSP 122 on the control of post-weaning diarrhea syndrome of piglets. Research Veterinary Science 67223228

30. M. Lessard, G. J. Brisson, 1987Effect of Lactobacillus Fermentation Product on growth immume response and fecal enzyme activity in weaned pigs. Can. J. Anim. Sci. 67509516

31. C. B. Maassen, C. van Holten-Neelen, F. Balk, Bak. den-Glashouwer, M. J. , R. J. Leer, J. D. Laman, W. J. Boersma, E. Claassen, 2000Strain-dependent induction of cytokine profiles in the gut by orally administered Lactobacillus strains. Vaccine1826132623

32. D. R. Mack, S. Michael, S. Wei, L. Mc Dougall, M. A. Holligsworth, 1999Probiotics inhibits enteropathogenic E. coli adherence in vitro by inducing intestinal mucin gene expression. American Journal of PhysiologyG941G950.

33. P. Mangell, P. Nejdfors, M. Wang, S. Ahrne, B. Westrom, I. Thorlacius, B. Jeppsson, 2002Lactobacillus plantarum 299v inhibits Escherichia coli induced intestinal permeability. Digestive Disease Science 47511516

34. K. Männer, A. Jadamus, W. Vahjen, U. Frackenpohl, O. Simon, 2002Effekte probiotischer Zusätze auf Leistungsparameter und intestinale Mikroflora. Proc. 7. Tagung, Schweine und Geflügelernährung;7880

35. K. Männer, A. Spieler, 1997Probiotics in piglets, an alternative to traditional growth promoters.Microecology and Therapy 26243256

36. M. L. Marin, J. H. Lee, J. Murtha, Z. Ustunol, J. J. Petka, 1997Differential cytokine production in clonal macrophage and T-cell lines cultured with bifidobacteria.Journal of Dairy Science8027132720

37. Vance. J. Mc Craken, H. Gaskins, Rex, 1999Probiotics and the immune system. Probiotics: A critical review. 85111

38. E. Metchnikoff, 1907The prolongation of lifeHeinemann, London, UK.

39. M. C. Moreau, G. Corthier, 1988Effect of gastrointestinal microflora on induction and maintenance of oral tolerance to ovalbumin in C3H/HeJ mice.Infection and Immunity, 5627662768

40. L. Morelli, 1995Variations of Coliformes and Lactobacilli spp content in liquid or soft faeces belonging at swines treaties and not treaties. Published by AKRON-firm, Milano. on the mucosal IgA response of mice to dietary antigens. Bioscience Biotechnology Biochemistry 621015

41. Neutra M.R, Forstner J.F (1987) Gastrointestinal mucus: Synthesis, secretion and function 975-1009. In L.R. Johnson (ed), Physiology of the gastrointestinal tract, 2nd edition. Raven Press, New York.

42. Oullette A.J, Selsted M.E1996Paneth cell defensins: Endogenous peptide components of intestinal host defense. FASEB journal 1012801289

43. G. Perdigon, S. Alvarez, P. de Ruiz, A. Holgado, 1991Immunoadjuvant activity of oral Lactobacillus casei: influence of dose on the secretory immune response and protective capacity in intestinal infections.Journal of Dairy Research 58485496

44. Pollmann D.S, Danielson D.M, Peo E.R1980Effects of microbial feed additives on performace of starter and growing-finishing pigs. J. Anim. Sci. 513577581

45. Pouwels P.H, Leer R.J, Boersma W.J1996The potential of Lactobacillus as a carrier for oral immunization: development

and preliminary characterization of vector systems for targeted delivery of antigens.J. Biotechnol. 44183192

46. Qu X.D, Lloyd K.C, Walsh J.H, Lehrer R.I1996Secretion of type II phospholipase A and cryptdin by rat small intestinal Paneth cells.Infect. Immun. 6451615165

47. Rolfe D.R2000The role of probiotic cultures in the control of gastrointestinal health.Symposium: Probiotic bacteria: Implications for Human Health.

48. R. Rolfe, 1996Colonization resistance. In R I Mackie, B A White and R E Isaacson (ed), Gastrointestinal microbiology. Gastrointestinal microbes and host interactions. Chapman and Hall. New York. 2501536

49. D. C. Savage, 1986Gastrointestinal micro flora in mammalian nutrition. Ann. Rev. Nutr, 6155178

50. L. Scharek, L. Hartmann, R. L. Heinevette, M. Blaut, 2000Bifidobacterium adolescentis modulates the specific immune response to another human gut bacterium, Bacteroides thetaiotaomicron, in gnotobiotic rats.Immunobiology202429441

51. H. Scumm, R. Pohl, H. Willeke, 1990Ergebnisse des Einsatzes von Suiferm bei Absatzferkeln mit Durchfällen zur Aufrechterhaltung und Wiederherstellung der gesunden Darmflora. Tierärztliche Umschau 45402411

52. Q. Shu, F. Qu, H. S. Gill, 2001Probiotic treatment using Bifidobacterium lactis HNO19 reduces weanling diarrhea associated lymphoid tissue in neonatal swine. Immunological Methods 241185199

53. O. Simon, A. Jadamus, W. Vahjen, 2001Probiotic feed additives-effectiveness and expected modes of action.J. Anim. Sci. 10 Supp 115167

54. Simon O, Vahjen W, Scharek L (2003) Microorganisms as Feed Additive-Probiotics. Proc. 9 th International Symposium on Digestive Physiology in Pigs, Banff, Canada; Vol 1: 295-318.

55. G. I. Solano-Aguilar, K. G. Vengroski, E. Beshah, J. K. Lunney, 2000Isolation and purification of lymphocyte subset from gut-associated lymphoid tissue in neonatal swine. Immunological Methods 241185199

56. C. R. Stokes, M. Bailey, A. D. Wilson, 1994Immunolofy of the porcine gastrointestinal tract. Veterinary Immunology and Immunopathology 43143150

57. T. Takahashi, E. Nakagawa, T. Nara, T. Yajima, T. Kuwata, 1998Effects of orally ingested Bifidobacterium longum

58. Tortuer OF1973Influence of implantation of Lactobacillus acidophilus in chicks on the growth, feed conversion, malabsorption of fats syndrome and intestinal flora. Poult. Sci. 2197203

59. Tournut J (1989) Les probiotique en élevage: applications. Revue Scientifique et Technique de l,Office International des Epizooties, 8: 533-549.

60. J. Travnicek, L. Mandel, I. Trebichavsky, M. Talafantova, 1989Immunological state of adult germfree miniature Minnesota pigs.Folia Microbiologia. 34157164

61. D. Tuschy, 1986Verwendung von' Probiotika' als Leistungsförderer in der Tierernährung. Übersichten Tierernährung 14157178

62. M. A. Vega-Lopez, G. Arenas-Contreras, M. Bailey, S. Gonzalez-Pozos, C. R. Stokes, M. G. Ortega, R. Mondragon-Flores, 2001Development of intraepithelial cells in the porcine small intestine. Dev Immunology 8147158

63. E. Veizaj, M. Tafaj, K. Männer, 2008The effect of the combined probiotic preparation on growth performance, digestibility and microbial composition of faeces of weaned piglets. Agricultura53741

64. E. Veizaj-Delia, Th. Piu, P. Lekaj, M. Tafaj, 2010Using combined probiotic to improve growth performance of weaned piglets on extensive farm conditionsLivestock Science1-3249 EOF251 EOF

65. E. Vitini, S. Alvarez, M. Medina, M. Medici, M. V. de Budeguer, G. Perdigon, 2001Gut mucosal immunostimulation by lactic acid bacteria. Biocell 24223232

66. C. Winckler, B. Schröder, G. Breves, 1998Effects of Saccharomyces boulardii, Bacillus cereus var. caron and Bacillus cereus var. toyoi on epithelial transport functions in pig jejunum. Zeitschrift für Gastroenterologiy (Suppl. 1) 30-37.

67. Wostmann B.S1996Nutrition. In B.S. Wostmann (ed), Germfree and gnotobiotic animal models. CRC Press, Boca Raton Fl, 7187

68. J. K. Zani, da. Weykamp, F. Cruz, dos. Freitas, A. Santos, C. Gil-Turnes, 1998Effect of probiotic CenBiot on the control of diarrhoea and feed efficiency in pigs.Journal of Applied Microbiology846871

69. G. Zimmerman, R. Bollinger, J. C. Mc Donald, 1970Possible production of immunosuppression in the host by exposure to microorganisms: observations in germfree rats.Surgery Forum. 21236238

CHAPTER 7

Appraisal of Chicken Production with Associated Biosecurity Practices in Commercial Poultry Farms Located in Jos, Nigeria

C. V. Maduka, I. O. Igbokwe, and N. N. Atsanda

Faculty of Veterinary Medicine, University of Maiduguri, Maiduguri 600001, Nigeria

ABSTRACT

A questionnaire-based study of chicken production system with on-farm biosecurity practices was carried out in commercial poultry farms located in Jos, Nigeria. Commercial and semicommercial farms had 75.3% and 24.5% of 95,393 birds on 80 farms, respectively. Farms using deep litter and battery cage systems were 69 (86.3%) and 10 (12.5%), respectively. In our biosecurity scoring system, a correct practice of each indicator of an event scored 1.00 and biosecurity score (BS) of each farm was the average of the scores of biosecurity indicators for the farm, giving BS of zero and 1.00 as absence of biosecurity and optimal biosecurity, respectively. Semicommercial farms had higher BS than commercial farms. The flock size did not significantly ()

affect the mean BS. Disease outbreaks correlated () with BS, showing a tendency of reduction of disease outbreaks with increasing BS. Outbreaks were significantly () associated with deep litter system. In conclusion, the chicken production system requires increased drive for excellent biosecurity practices and weak points in the biosecurity could be ameliorated by extension of information to farmers in order to support expansion of chicken production with robust biosecurity measures that drastically reduce risk of disease outbreak.

1. INTRODUCTION

Entrepreneurial initiatives in the commercial poultry (chicken) industry increased in recent time in Jos, Nigeria, in response to national animal protein demand and as a result of the mild climatic environment and apparent profitability of the business. The industry is characterized by low to medium integration of inputs representing the downstream poultry sector where the flock stocking is low and involves raising day-old chicks, supplied by integrated farms, to mature weight (broilers) or egg-laying (laying hens) using commercial feeds and veterinary products from retail outlets [1, 2]. The chickens are fully housed in intensive management system and often reared on deep litter within urban and periurban locations. The production system is oriented towards profitability and the efforts to reduce cost of production may impact negatively on the system through neglect of important pivotal elements of poultry preventive health. Communicable diseases and problems related to feeding constitute the major constraint to profitable chicken production in the locality [2]. Therefore, disease prevention and control are veritable components of commercial poultry production. Biosecurity refers to principles engaged in reducing the chance of introduction and spread of pathogens within and between farms by preventing infectious agents from entering (bioexclusion) or exiting (biocontainment) the farm and the principal elements are segregation, traffic control, cleaning, and disinfection [3, 4]. An effective biosecurity has conceptual, structural, and operational frameworks which involve housing design and construction with management procedures that keep the flock free from infectious diseases [5–7]. The application of biosecurity is affected by small-holder, farm, social, and community characteristics [7, 8]. There have been reports of breaches in biosecurity measures in poultry production systems in parts of Nigeria because of lack of awareness and failure to implement components of biosecurity [9–11] resulting in frequent outbreaks of diseases which drastically reduce profit or lead to capital loss in the industry [2, 12]. The operational cost of biosecurity is usually low and there is a high benefit-cost ratio [4, 7, 13], but inadequate implementation of biosecurity measures may be due to insufficient motivation and lack of understanding of its economic benefits [14].

We hypothesized that the structure of chicken production in poultry farms, in the study area, was commercially oriented, family-based, and income-generating investment, which were set up as small-holder, farm-based intensive production system with some deficiencies in biosecurity practices and constraints associated with disease outbreaks [2]. The objective of this study, therefore, was to appraise chicken production along with the on-farm biosecurity practices (consistent with standard bioexclusion and biocontainment protocols) in commercial poultry farms located in Jos, Nigeria, and to assess whether some characteristics of the farms affect biosecurity rating and effectiveness in reducing outbreaks of communicable disease.

2. MATERIALS AND METHODS

2.1. Study Area and Poultry Farms

The study was conducted in Jos North and South Local Government Areas located between 9°56′N and 8°53′E in Plateau State, Nigeria (Figure 1). The locations of 80 poultry farms were identified in the area through veterinarian-clientele relationship.

2.2. Questionnaire Construction

A questionnaire was used to gather information from 80 poultry farms on the following:(a)Poultry farm production system. (b)Biosecurity events outside the premises, at the farm boundary, between the farm boundary and poultry house, and within the poultry house.(c)Outbreak of any communicable disease causing mortality after application of biosecurity in the current stocking cycle. The disease was syndromally diagnosed by the consulting veterinarian and included Marek disease, coccidiosis, infectious bursal disease, avian influenza, Newcastle disease, infectious coryza, fowl pox, fowl cholera, or egg drop syndrome after postmortem examination.The questionnaire was structured to obtain "yes" or "no" answers and only open for specific additional responses for clarity of answers.

2.3. Filling the Questionnaire

The questionnaire was filled by the farmer or farm personnel during farm visits conducted in February–July 2014 and validated by information obtained after observation of the farm environment, interview of the farmer, and checking of farm records on the farm and in the veterinary clinic where the farm was registered. The validation was necessary in order to eliminate response bias which could weaken the strength of questionnaire survey [15].

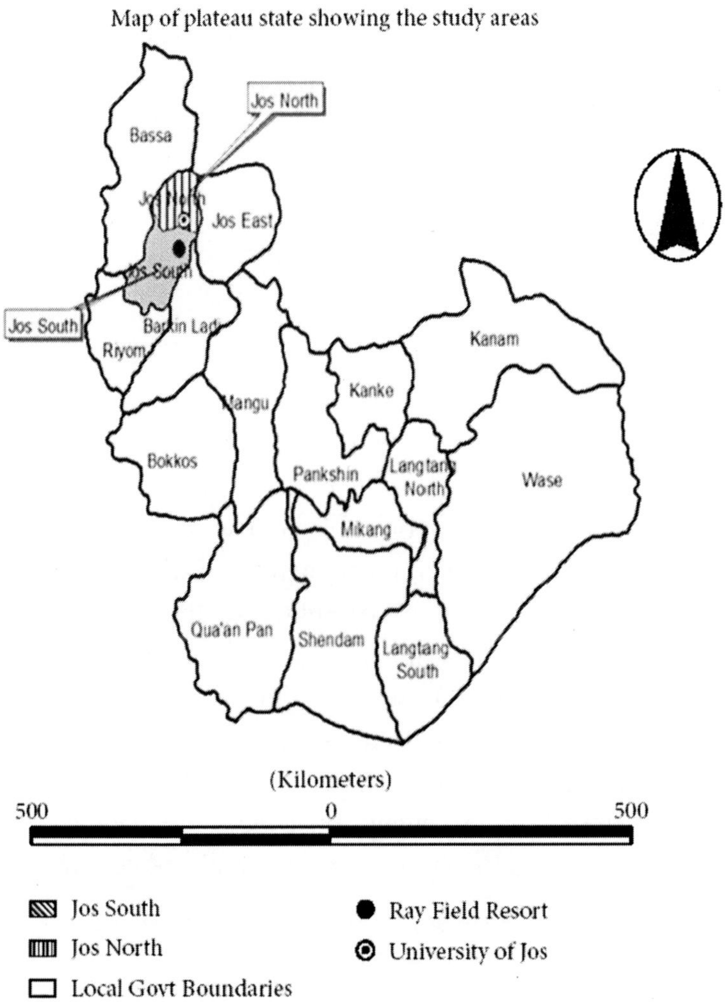

Figure 1. Map of Jos showing Jos North and South Local Government Areas among other Local Government Areas in Plateau State, Nigeria

2.4. Biosecurity Scoring System

A biosecurity scoring system was developed from the indicators of biosecurity events observed in the evaluation of biosecurity practices on the farm as previously reported [5, 8, 16–23]. The biosecurity indicators were 36 and the correct practice of each indicator was scored as 1.00. The biosecurity score (BS) of each farm was the average of the scores of the biosecurity indicators for the farm, giving a BS of zero and 1.00 as absence of biosecurity and optimal biosecurity, respectively.

2.5. Statistical Analysis

Data were collected from the questionnaire to obtain the number of positive responses for every item on the questionnaire as well as the number and ages of birds involved. The number of positive responses from respondents on each item was also presented as a percentage of total number of respondents and the response rate for the item was calculated as a percentage of the number of farms administered with questionnaires. The BS of farms were summarised as means and standard deviations. The associations and relationships were assessed using Chi-square and Pearson's correlation, respectively; and variations in means were assessed by one-way ANOVA with Tukey post hoc test using computer software (GraphPadInstat:http://www.graphpad.com/apps/index.cfm).

3. RESULTS

3.1. Chicken Production System

The number of farms and chickens involved with associated reasons for poultry production, type of poultry, flock size, age of birds, housing system, sources of day-old chicks (DOC), and feed are presented in Tables 1(a) and 1(b). Very few farms (3.7%) produced chickens for family use only (consumption), but many farms (37.5%) produced for family use and marketed the surplus. The highest proportion of farms (58.8%) produced for strictly commercial purposes. Farms made incomes from sale of live chickens and table eggs to wholesale buyers, manure (faecal droppings) to crop farmers, and dead birds to dog breeders and keepers. The number of birds on strict commercial farms was 75.3% of the total number (95,393) of birds on 80 farms; whereas 24.5% of the birds were on semicommercial farms. Fifty-one (63.5%) farms raised only layers while 17 (21.2%) farms had both broilers and layers. Cockerel-layer and broiler farms were 1 (1.2%) and 11 (13.8%), respectively. The chicken population consisted of layers (48.8%), broilers (3.9%), broilers and layers (46.8%), or cockerels and layers (0.5%). Farms with flock size of 200–500 birds (36.2%) were most common followed by those with 501–1000 (25.0%) and 1001–2000 (18.8%) birds. The flock sizes with 30.7%, 25.3%, 16.5%, and 15.7% of the total bird population (TBP) were >10,000, 1,001–2,000, 501–1,000, and 2,000–10,000 birds per farm, respectively. Most birds (69.7%) were ≥18 weeks old in 54 (67.5%) farms while those that were <13 weeks old (28.3%) were in 23 (28.8%) farms. The older birds were usually layers and the younger ones were either broilers or growing layers. The farms using deep litter system were 69 (86.3%) housing 77.1% of TBP. Battery cages were used solely in 10 (12.5%) farms or in combination with deep litter in 1 (1.2%) farm with 7.9% and 15.0% of TBP, respectively. The day-old chicks, reared in 28

(35.0%) and 16 (20.0%) farms having 39.1% and 32.1% of TBP, were sourced from Zartech and Obasanjo farms, respectively. Few farms (–7, 2.5–8.7%) had other sources of DOC consisting of Agrited, ECWA, Amo, and Chi farms. Feeds used were commercial type in 68 (85%) farms housing 86.9% of TBP. The frequently used brands were Livestock, Vital, and Hybrid feeds in 19 (29.4%), 18 (26.5%), and 16 (23.5%) farms housing 41.4%, 14.9%, and 12.8% of TBP.

3.2. Biosecurity Appraisal

The frequencies of positive responses on indicators of biosecurity events in chicken-producing poultry farms with validated responses on a structured questionnaire are in Table 2. The biosecurity indicators that had >90% positive responses included presence of good storage facility, appropriate carcass disposal, rodent-proof facility, separation of poultry types and birds of varying ages, proper ventilation and availability of clean water, appropriate and dry bedding, washing and disinfection of poultry houses prior to restocking, regular washing of feeders and drinkers, and isolation of sick birds. The positive responses dropped to 80–90% in terms of awareness of the scope of appropriate biosecurity practices, fencing with gate, use of only certified commercial feed, washing hand and showering before and after handling birds, frequent changing of bedding, and practice of all-in all-out management. The positive responses further dropped to 60–80% regarding stocking DOC from certified sources, regularly disinfecting feeders and drinkers, use of functional footbath at the entrance of the poultry house, chemoprophylactic treatments of apparently healthy birds, and consulting veterinarians in the event of health challenges on the farm. Positive responses were 50–60% in the aspects of providing parking lot outside the farm premises and functional footbath at farm boundaries. Positive responses decreased to <50% in appropriate biosecurity practices of washing and disinfecting vehicles that drive into farm premises and residence of farm workers within the premises. Some farmers (47.4%) allowed visitors into the poultry premises. Positive responses for inappropriate biosecurity practices were <32% in situations involving clustering of farms (26.6%), water bodies for migratory birds in the neighbourhood (20.2%), acquisition of second-hand equipment (6.3%), on-farm carnivorous pets (8.9%), necropsies (31.2%), and allowing of birds to occasionally move out of their pens (3.8%).

Biosecurity scoring of 63 farms, with ≥34 validated responses related to biosecurity, showed that the farms had a mean BS of with minimum and maximum scores of 0.50 and 0.94, respectively. Table 3 presents the effects of reason for keeping poultry and flock size on biosecurity scores. Mean BS was significantly () affected by reason for poultry production, with semicommercial farms having higher BS than strictly commercial farms and noncommercial farms having comparable mean BS () with strictly commercial farms. The flock size did not influence the mean BS (), but the highest mean BS was in farms with >10,000 birds per farm. Table 4 presents frequencies of farms with various classes of BS in relation to occurrence of disease outbreak. The farms classified with good (0.50–0.70), very good (0.71–0.90), and excellent (0.91–1.00) BS consisted of 11 (17.5%), 45 (71.4%), and 7 (11.1%) farms, respectively. Disease outbreaks negatively correlated () with BS showing that there was a tendency of reduction of disease outbreaks with increasing BS. Outbreaks occurred in 38 (55.1%) out of 69 farms using deep litter and no outbreak occurred in 10 farms using battery cages (see Table 5). There was significant () association of outbreak with deep litter system.

Table 1. (a) Number (%) of farms surveyed and birds involved with various reasons for farming, poultry type, stock size, and ages of birds. (b) Number (%) of farms and birds with various housing systems and sources of day-old chicks and feed.

(a)

	Number (%) of farms	Number (%) of birds
Reason for keeping birds		
Commercial	47 (58.8)	71843 (75.3)
Family use (noncommercial)	3 (3.7)	150 (0.2)
Both (semicommercial)	30 (37.5)	23400 (24.5)
Total	80 (100.0)	95393 (100.0)
Poultry type		
Broiler	11 (13.8)	3701 (3.9)
Layer	51 (63.8)	46540 (48.8)
Both broiler and layer	17 (21.2)	44672 (46.8)
Cockerel and layer	1 (1.2)	480 (0.5)
Total	80 (100.0)	95393 (100.0)
Number of birds per farm		
<200	9 (11.2)	743 (0.8)
200–500	29 (36.2)	10460 (11.0)
501–1000	20 (25.0)	15795 (16.5)
1001–2000	15 (18.8)	24145 (25.3)
2000–10000	5 (6.3)	14950 (15.7)
>10000	2 (2.5)	29300 (30.7)
Total	80 (100.0)	95393 (100.0)
Age of birds at study time		
<13 wks	23 (28.8)	26966 (28.3)
13–17 wks	3 (3.7)	1900 (2.0)
≥18 wks	54 (67.5)	66527 (69.7)
Total	80 (100.0)	95393 (100.0)

(b)

	Number (%) of farms	Number (%) of birds
Housing system		
Deep litter	69 (86.3)	73545 (77.1)
Battery cage	10 (12.5)	7548 (7.9)
Both	1 (1.2)	14300 (15.0)
Total	80 (100.0)	95393 (100.0)
Source of day-old chicks		
Obasanjo	16 (20.0)	30616 (32.1)
Zartech	28 (35.0)	37302 (39.1)
Zartech and Agrited	2 (2.5)	4500 (4.7)
Agrited	5 (6.3)	3350 (3.5)
ECWA	6 (7.5)	5398 (5.7)
Amo	6 (7.5)	4280 (4.5)
Chi	7 (8.7)	4600 (4.8)
Unknown	10 (12.5)	5347 (5.6)
Total	80 (100.0)	95393 (100.0)
Sources of feed		
Commercial	68 (85.0)	82896 (86.9)
Vital	18 (26.5)	12370 (14.9)
Livestock	19 (29.4)	34307 (41.4)
Amo	6 (8.8)	4242 (5.1)
Top	2 (2.9)	680 (0.8)
Hybrid	16 (23.5)	10622 (12.8)
Vital and top	1 (1.5)	15000 (18.1)
Unknown	5 (7.4)	5675 (6.9)
On-farm (local) preparation	6 (7.5)	6850 (7.1)
Both	6 (7.5)	5647 (6.0)
Total	80 (100.0)	95393 (100.0)

Table 2. Frequency of biosecurity events on poultry farms in Jos, Nigeria.

Indicators of biosecurity events	Number (%) of responses	Number (%) of farms with "yes" response
Events outside the premises		
Awareness of biosecurity practices	79 (98.8)	67 (84.8)
Density of (>5) farmers in neighbourhood	79 (98.8)	21 (26.6)
Water bodies for migratory birds in neighbourhood	79 (98.8)	16 (20.2)
Certified sources of quality chicks	76 (95.0)	52 (68.4)
Parking lot outside the farm premises	77 (96.3)	42 (54.5)
Acquisition of second-hand equipment	79 (98.8)	5 (6.3)
Farm boundary events		
Fencing with gates	78 (97.5)	66 (84.6)
Washing/disinfection of vehicles	76 (95.0)	17 (22.3)
Functional[†] footbath at entry point	78 (97.5)	44 (56.4)
Visitors allowed into premises	78 (97.5)	37 (47.4)
Events between farm boundary and poultry house		
Presence of good feed storage facility	79 (98.8)	75 (94.9)
Appropriate carcass disposal	80 (100.0)	78 (97.5)
On-farm necropsy	79 (98.8)	25 (31.6)
Certified commercial feed sources only	80 (100.0)	68 (85.0)
On-farm carnivores (dogs and cats)	79 (98.8)	7 (8.9)
Hand washing/shower before and after handling birds	79 (98.8)	65 (82.3)
Rodent-proof	78 (97.5)	74 (94.9)
Residence of farm workers within premises	79 (98.8)	39 (49.4)
Functional[†] footbath at entrance of poultry house	77 (96.3)	56 (72.7)

Events inside poultry house		
Separation of poultry types	80 (100.0)	78 (97.5)
Separation of birds according to age	80 (100.0)	79 (98.8)
Proper ventilation	80 (100.0)	79 (98.8)
Availability of clean water	77 (96.3)	76 (98.7)
Appropriate bedding material	69 (86.3)	66 (95.7)
Dry bedding	65 (81.3)	64 (98.5)
Frequent changing of bedding	66 (82.5)	59 (89.4)
Birds occasionally allowed to move out of the poultry house	78 (97.5)	3 (3.8)
Washing/disinfecting poultry house prior to restocking	78 (97.5)	76 (97.4)
Practice of all-in all-out management system	77 (96.3)	69 (89.6)
Washing feeders/drinkers regularly	78 (97.5)	77 (98.7)
Disinfecting feeders/drinkers regularly	75 (93.8)	48 (64.0)
Isolation of apparently sick birds	76 (95.0)	74 (97.4)
Prophylactic chemotherapy to apparently healthy birds	77 (96.3)	57 (74.0)
Consultation of veterinarians only in the event of problems	78 (97.5)	60 (76.9)

[+]Functional footbath comprises (a) presence of the footbath and (b) frequent (daily or once every two days) replenishment of the same.

Table 3. Effects of reasons for keeping poultry and number of birds per farm on biosecurity scores.

	Biosecurity score* (number of farms‡)
Reason for keeping poultry	
Commercial	0.75 ± 0.10^{a} (32)
Family use only (noncommercial)	0.82 ± 0.07^{ab} (3)
Both (semicommercial)	0.87 ± 0.05^{b} (28)
Number of birds per farm	
<200	0.80 ± 0.07^{a} (7)
200–500	0.81 ± 0.11^{a} (23)
501–1000	0.79 ± 0.13^{a} (15)
1001–2000	0.80 ± 0.10^{a} (12)
2001–10000	0.85 ± 0.03^{a} (4)
>10000	0.88 ± 0.02^{a} (2)

4. DISCUSSION

This study has provided the first expanded data base on the structure of chicken production system in Jos, Nigeria. Although a strictly commercial farm-based approach is the mainstay of the production system, a large contribution to the production capacity was made from family production which was usually semicommercial. Earlier reports in Nigeria indicated the preponderance of small-scale family chicken production as a means of reducing food insecurity and improving family income [2, 9, 18, 24–27]. The flock size per farm was frequently <1000 with the commonest flock size of 200–500, but aggregate population of chickens in this type of holding was just 38.3% of TBP. Previous reports, in the locality, indicated the most frequent flock size to be <200 [2] and <100 [26]. Family farms in other parts of Nigeria were reported to mostly have flock sizes of <200 [9] and ≤500 [27]. The production capacity increased when the flock sizes were >1000 such that the contribution to poultry sector was greater with larger than smaller farms. Thus, efficiency of production was expected to increase in larger farms as regards measures to mitigate production losses.

In our study, there were more layer than broiler farms and more layers than broilers in the TBP. Muhammad et al. [26] reported more broiler than layer farms, but more layers than broilers in stocking capacity. The larger population of birds, at the age of egg-laying compared to other ages, indicated that layers were more numerous in the bird population. Most of the birds were housed on deep litter as similarly reported by others [2, 9, 26]. More than 70% of DOC were obtained from two hatcheries and used by 55% of the farms indicating that the DOC suppliers probably had larger supply capacity outlet in Jos. The feeds were sourced from commercial feed producers by 85% of farms with 86.9% of TBP and three feed firms had the largest market perhaps because of logistic advantages. Therefore, farmers prefer commercial feeds to on-farm feed preparation because of, probably, better feed efficiency and cost effectiveness with commercial than on-farm feeds. Other DOC and feed suppliers are vital to the supply chain and may be gaining more market space as the production system expands.

The level of awareness of biosecurity practices among poultry farmers in Jos was high and as a result, no farm had poor BS of <0.5. Rather, some farms had good or excellent BS and most of them had very good BS. The awareness of the need for biosecurity is often elicited by veterinary advice which comes along with veterinary services during disease outbreaks [28]. Inadequate understanding of the scope of biosecurity practice remains the hindrance to biosecurity compliance in parts of northeastern Nigeria [11], but adequate biosecurity measures were reported in commercial poultry farms in the same region while some backyard farms lacked such measures [9]. In Egypt, biosecurity measures were rarely implemented in small-scale commercial production units [29]. In Jos, most small-scale farms returned positive responses (>60%) on several indicators of important biosecurity events. Deficiency in biosecurity compliance was remarkable when positive responses dropped to <60% in aspects of providing parking lot outside the farm and functional footbath at farm boundaries. Furthermore, vehicles drove into farm premises without washing and disinfection; farm workers had residences outside the farm premises in >50% of farms and some farms used second-hand equipment like egg crates. It was necessary to avoid these areas of biosecurity failures [4]. The risks associated with avian influenza outbreaks in Nigeria were receiving visitors to the farm, purchasing of live poultry or products, and workers living outside the farm premises [10, 30, 31]. Therefore, biosecurity compliance is a compact practice which should not give room for any gaps in the production system for entry or exit of infectious agents [7].

The biosecurity scoring system adopted in this report incorporated a broad scope of indicators of a variety of biosecurity events, thereby making it comprehensive for the local production system. Our biosecurity scoring system scored each indicator equally in a positive or negative sense [21] and offered an average BS for a farm but differed from other scoring systems with scores of 0–3 [22] or 1–3 [8, 20] for an indicator. The current scoring system provided the quantitative means to determine the extent production system and flock size influenced the biosecurity scores of farms. Semicommercial farms had significantly higher BS than strictly commercial farms; and it is presumed that employed labour in strictly commercial farms may be nonchalant about aspects of biosecurity in the absence of owners' supervision and appropriate motivation. On-farm biosecurity was reported to be affected by the level of motivation derived from the information on costs and accrued benefits [14]. The flock size did not significantly influence BS, but highest mean BS was in farms with the largest flock size. The contribution to the production capacity of the poultry sector was greater with larger compared to smaller farms; and as a result, enhanced efficiency of the production system was expected as regards investments in biosecurity measures to mitigate production losses. Dorea et al. [28] reported that flock size did not affect the standard of biosecurity protocols implemented, maybe because the awareness of biosecurity was broad and cut across the spectrum of small and large scale farm managements as the case was in this study. However, other reports indicated that farmers with larger farm areas and flock sizes tended to have enhanced biosecurity practices [8, 13].

Occurrence of outbreaks of communicable disease on farms decreased with increasing BS, thereby, affirming the importance of biosecurity in the production system's disease control. There have been reports of reduction of infectious disease outbreaks with standard biosecurity protocols [6, 32]. Biosecurity failures seemed to be a greater challenge in deep litter than battery cage farms, since outbreaks were significantly associated with deep litter system. The need for disease control in deep litter is amplified by the fact that farmers rarely use battery cage because of cost. Disease transmission occurs on deep litter by contact through inhalation of aerosolized particles and ingestion of contaminated feed and water [33–36]. In order to prevent infections, farmers (74% of respondents) engage in prophylactic treatments of apparently healthy birds, sometimes without proper veterinary supervision and with adverse consequences [37, 38]. Development of antimicrobial resistance may occur as a result of this practice [38] and antimicrobial residues have been reported in poultry meat and eggs when drug withdrawal period is not observed with the implication of food chain contaminations [39, 40].

Table 4. Number (%) of farms with various scores on biosecurity in relation to occurrence of disease outbreak#.

Biosecurity score*	Number (%) of farms	
	All responding farms‡	Farms with disease outbreak
<0.5	0	0
0.50–0.70	11 (17.5)	9 (81.8)
0.71–0.90	45 (71.4)	24 (53.3)
0.91–1.00	7 (11.1)	3 (42.9)
<0.5–1.00	63 (100)	36 (57.1)

* Average score on correct responses on biosecurity indicators on a farm, where incorrect and correct scores are 0 and 1, respectively.

‡ Farms with ≥ 34 responses on biosecurity.

Inverse relationship between biosecurity scores in first column with percentages of disease outbreak in third column is computed to give a correlation coefficient of −0.97.

Table 5. Effect of housing system on the occurrence of disease outbreaks on farms.

Housing system	Number (%) of farms	
	All responding farms	Farms with disease outbreak
Deep litter	69 (86.3)	38 (55.1)[a]
Battery cage	10 (12.5)	0 (0)[b]
Both (deep litter and battery cage)	1 (1.2)	1 (100.0)[ab]
Total	80 (100.0)	39 (48.8)

[a,b]Unmatched superscripts indicate significant ($p < 0.05$) difference in the column.

All the farms were registered for veterinary services and most of them sought for such services only in the event of health care challenges. Regular veterinary farm visits through retainership programs was not often in practice. Farmers call for veterinary investigation of mortality of chickens and necropsies are usually carried out as part of the diagnostic process. Ideally, the carcasses are moved in a biosecure manner to veterinary diagnostic facilities. Some farms, in this study, reported carrying out necropsies on the farm and this was considered a negative biosecurity event especially where disinfection and appropriate disposal of carcasses were not ascertained. Most farmers in Jos sold dead chickens which were subsequently dressed and cooked for consumption by dogs. This practice is considered to have biosecurity implications and was presumed to be an effort to reduce financial loss. By scouting for dog owners who have established demand windows, farmers enhance the value chain in the industry. Carcasses of dead chicken on poultry farms were reported to be disposed by incineration, deep burial, or dumping as refuse [9]. Dumping of carcasses was rare and farmers knew the associated biosecurity risks of such act. Few farmers had on-farm carnivorous pets (dogs and cats) and were not aware of negative biosecurity implications of the pets having possible chance of eating unprocessed dead carcasses.

In conclusion, the questionnaire-based study revealed that the intensive chicken production system in Jos lacked integration of all inputs and consisted mostly of small farms, but few larger farms had aggregate bird population exceeding those of small farms. Semicommercial farms had higher BS than strictly commercial farms. Outbreaks of communicable disease negatively correlated with BS in deep litter system, and farmers engaged in prophylactic treatments to reduce production loses that might arise from biosecurity failures. The weak points in the biosecurity were identified and could be strengthened by providing the farmers information and focus group training to fill the gap in understanding of the scope of engagement of biosecurity practice as an important phenomenon in preventive poultry health.

COMPETING INTERESTS

The authors declare that they have no competing interests.

ACKNOWLEDGMENTS

The authors express their sincere thanks to the poultry farmers who responded to the questionnaires, Dr. C. C. Onah of Health and Agricultural Technical Services (HATS Vet), Jos, Nigeria, who provided the connections to all the poultry farms through clientele relationships, and Mr. Sylvester Ogbaji who coordinated the field work. C. V. Maduka provided private funds for logistics during data collection for his DVM dissertation submitted to the Faculty of Veterinary Medicine, University of Maiduguri, Maiduguri, Nigeria.

REFERENCES

1. D. F. Adene and A. E. Oguntade, Poultry Sector Country Review, Nigeria, Food and Agriculture Organization, Rome, Italy, 2006.

2. U. Haruna, S. A. Jibril, D. J. U. Kalla, and H. Suleiman, “Evaluation of egg production in Jos North Local Government

Area, Plateau State, Nigeria," International Journal of Poultry Science, vol. 6, no. 8, pp. 604–607, 2007.

3. N. Charisis, "Avian Influenza biosecurity: a key for animal and human protection," Veterinaria Italiana, vol. 44, no. 4, pp. 657–669, 2008.

4. F. O. Fasina, A. M. Ali, J. M. Yilma, O. Thieme, and P. Ankers, "The cost-benefit of biosecurity measures on infectious diseases in the Egyptian household poultry," Preventive Veterinary Medicine, vol. 103, no. 2-3, pp. 178–191, 2012.

5. S. M. Shane, Handbook of Poultry Diseases, American Soybean Association, Singapore, 2nd edition, 2005.

6. D. A. Halvorson, "Biosecurity on a multiple-age egg production complex: a 15-year experience," Avian Diseases, vol. 55, no. 1, pp. 139–142, 2011.

7. K.-M. Siekkinen, J. Heikkilä, N. Tammiranta, and H. Rosengren, "Measuring the costs of biosecurity on poultry farms: a case study in broiler production in Finland," Acta Veterinaria Scandinavica, vol. 54, article 12, 2012.

8. S. H. Susilowati, I. Patrick, M. Iqbal, and T. Jubb, "The characteristics of the farm and the farmer that affect the adoption of biosecurity on small holder poultry farms in Indonesia," Livestock Research for Rural Development, vol. 25, no. 5, article 88, 2013,

9. Y. A. Geidam, H. I. Gambo, S. B. Adamu, H. A. Grema, A. M. Dapchi, and K. A. Sanda, "An assessment of the biosecurity measures in poultry farms in Borno and Yobe States," Sahel Journal of Veterinary Science, vol. 10, no. 2, pp. 83–86, 2011.

10. A. M. Wakawa, S. B. Oladele, P. A. Abdu, L. Sa'idu, and S. B. Mohammed, "Risk factors for the occurrence and spread of highly pathogenic avian influenza H5N1 in commercial poultry farms in Kano, Nigeria," Sokoto Journal of Veterinary Sciences, vol. 10, no. 2, pp. 40–51, 2012.

11. C. Augustine, D. I. Mojaba, J. S. Neils, and Y. U. Ngiki, "Assessment of factors affecting the implementation of biosecurity protocols by poultry farmers in Mubi area of Adamawa State, Nigeria,"International Journal of Management and Social Sciences Research, vol. 3, no. 2, pp. 13–16, 2014.

12. M. A. Maikasuwa and M. S. M. Jabo, "Profitability of backyard poultry farming in sokoto metropolis, Sokoto State, North-West, Nigeria," Nigerian Journal of Basic and Applied Sciences, vol. 19, no. 1, pp. 111–115, 2011.

13. O. K. Akintunde and A. I. Adeoti, "Assessment of factors affecting the level of poultry disease management in Southwest, Nigeria," Trends in Agricultural Economics, vol. 7, no. 2, pp. 41–56, 2014.

14. M. Laanen, D. Maes, C. Hendriksen et al., "Pig, cattle and poultry farmers with a known interest in research have comparable perspectives on disease prevention and on-farm biosecurity," Preventive Veterinary Medicine, vol. 115, no. 1-2, pp. 1–9, 2014.

15. R. Nespeca, J.-P. Vaillancourt, and W. E. Morgan Morrow, "Validation of a poultry biosecurity survey,"Preventive Veterinary Medicine, vol. 31, no. 1-2, pp. 73–86, 1997.

16. P. Nyaga, Strategies for the Prevention and Control Of Infectious Diseases (including Highly Pathogenic Avian Influenza in East Africa). Good Biosecurity Practices in Small

Scale Commercial and Scavenging Production Systems in Kenya, FAO, Rome, Italy, 2007.

17. O. Thieme, "Trends, issues and opinions in applying long term biosecurity measures on production systems and sector structure," in Proceedings of the Technical Meeting on Highly Pathogenic Avian Influenza and Human H5N1 Infection (HPAI '07), FAO, Rome, Italy, June 2007,

18. T. U. Obi, A. Olubukola, and G. A. Maina, "Pro-poor HPAI risk reduction strategies in Nigeria,"Africa/Indonesia Region Report 5, International Food Policy Research Institute (IFPRI) and Royal Veterinary College (RVC), 2008,

19. A. Conan, F. L. Goutard, S. Sorn, and S. Vong, "Biosecurity measures for backyard poultry in developing countries: a systematic review," BMC Veterinary Research, vol. 8, article 240, 2012.

20. V. S. Lestari, S. N. Sirajuddin, and K. Kasim, "Adoption of biosecurity measures by layer small holders,"Journal of the Indonesian Tropical Animal Agriculture, vol. 36, no. 4, pp. 297–302, 2011.

21. S. Van Steenwinkel, S. Ribbens, E. Ducheyne, E. Goossens, and J. Dewulf, "Assessing biosecurity practices, movements and densities of poultry sites across Belgium, resulting in different farm risk-groups for infectious disease introduction and spread," Preventive Veterinary Medicine, vol. 98, no. 4, pp. 259–270, 2011.

22. H. Wei and W. Aengwanich, "Biosecurity evaluation of poultry production cluster (PPCs) in Thailand,"International Journal of Poultry Science, vol. 11, no. 9, pp. 582–588, 2012.

23. E. Martindah, N. Ilham, and E. Basuno, "Biosecurity level of poultry cluster (PPC) in West Java, Indonesia," International Journal of Poultry Science, vol. 13, no. 7, pp. 408–415, 2014.

24. M. B. Abubakar, A. G. Ambali, and T. Tamjdo, "Rural chicken production: effects of gender on ownership, and management responsibilities in some parts of Nigeria and Cameroon," International Journal of Poultry Science, vol. 6, no. 6, pp. 413–416, 2007.

25. R. O. Babatunde, O. A. Omotesho, and O. S. Sholotan, "Socio-economics characteristics and food security status of farming households in Kwara State, North-Central Nigeria," Pakistan Journal of Nutrition, vol. 6, no. 1, pp. 49–58, 2007.

26. M. Muhammad, L. U. Muhammad, A. G. Ambali, and A. U. Mani, "A survey of early chick mortality on small-scale poultry farms in Jos, central Nigeria," International Journal of Poultry Science, vol. 9, no. 5, pp. 446–449, 2010.

27. N. S. Esiobu, G. C. Onubuogu, and V. B. N. Okoli, "Determinants of income from poultry egg production in Imo State, Nigeria: an econometric model approach," Global Advanced Research Journal of Agricultural Science, vol. 3, no. 7, pp. 187–199, 2014.

28. F. C. Dorea, R. Berghaus, C. Hofacre, and D. J. Cole, "Survey of biosecurity protocols and practices adopted by growers on commercial poultry farms in Georgia, USA," Avian Diseases, vol. 54, no. 3, pp. 1007–1015, 2010.

29. E. Negro-Calduch, S. Elfadaly, M. Tibbo, P. Ankers, and E. Bailey, "Assessment of biosecurity practices of small-scale broiler producers in central Egypt," Preventive Veterinary Medicine, vol. 110, no. 2, pp. 253–262, 2013.

30. F. O. Fasina, A. L. Rivas, S. P. R. Bisschop, A. J. Stegeman, and J. A. Hernandez, "Identification of risk factors associated with highly pathogenic avian influenza H5N1 virus infection in poultry farms, in Nigeria during the epidemic of 2006-2007," Preventive Veterinary Medicine, vol. 98, no. 2-3, pp. 204–208, 2011.

31. R. Métras, K. B. Stevens, P. Abdu et al., "Identification of potential risk factors associated with highly pathogenic avian influenza subtype H5N1 outbreak occurrence in Lagos and Kano States, Nigeria, during the 2006-2007 epidemics," Transboundary and Emerging Diseases, vol. 60, no. 1, pp. 87–96, 2013.

32. J. C. Gibbens, S. J. S. Pascoe, S. J. Evans, R. H. Davies, and A. R. Sayers, "A trial of biosecurity as a means to control Campylobacter infection of broiler chickens," Preventive Veterinary Medicine, vol. 48, no. 2, pp. 85–99, 2001.

33. M. Guarino, A. Caroli, and P. Navarotto, "Dust concentration and mortality distribution in an enclosed laying house," Transactions of the American Society of Agricultural Engineers, vol. 42, no. 4, pp. 1127–1133, 1999.

34. D. Weitzenbürger, A. Vits, H. Hamann, and O. Distl, "Effect of furnished small group housing systems and furnished cages on mortality and causes of death in two layer strains," British Poultry Science, vol. 46, no. 5, pp. 553–559, 2005.

35. T. B. Rodenburg, F. A. M. Tuyttens, K. De Reu, L. Herman, J. Zoons, and B. Sonck, "Welfare assessment of laying hens in furnished cages and non-cage systems: an on-farm comparison," Animal Welfare, vol. 17, no. 4, pp. 363–373, 2008.

36. J. de Vylder, S. van Hoorebeke, R. Ducatelle et al., "Effect of the housing system on shedding and colonization of gut and internal organs of laying hens with Salmonella Enteritidis," Poultry Science, vol. 88, no. 12, pp. 2491–2495, 2009.

37. I. O. Igbokwe, "Haemorrhagic syndrome due to sulphonamide toxicity," Zariya Veterinarian, vol. 3, p. 100, 1988.

38. J. R. Lawal, S. M. Jajere, Y. A. Geidam, A. M. Bello, Y. Wakil, and M. Mustapha, "Antiobiotic residues in edible poultry tissues and products in Nigeria: a potential public health hazard," International Journal of Animal and Veterinary Advances, vol. 7, no. 3, pp. 55–61, 2015.

39. J. Kabir, V. J. Umoh, E. Audu-okoh, J. U. Umoh, and J. K. P. Kwaga, "Veterinary drug use in poultry farms and determination of antimicrobial drug residues in commercial eggs and slaughtered chicken in Kaduna State, Nigeria," Food Control, vol. 15, no. 2, pp. 99–105, 2004.

40. J. J. Sasanya, J. W. O. Okenga, F. Ejobi, and M. Muganwa, "Use of sulfonamides in layers in Kampala district, Uganda and sulfonamide residues in commercial eggs," African Health Sciences, vol. 5, no. 1, pp. 33–39, 2005.

CHAPTER 8

Serological Survey for Avian Influenza in Turkeys in Three States of Southwest Nigeria

Daniel Oladimeji Oluwayelu, Comfort Oluladun Aiki-Raji, Oladunni Taiwo Adigun, Opeyemi Kazeem Olofintuyi, and Adebowale Idris Adebiyi

Department of Veterinary Microbiology and Parasitology, University of Ibadan, Ibadan 20005, Nigeria

ABSTRACT

Since the first outbreak of avian influenza (AI) in Nigeria in 2006, there has been continuous monitoring of the disease in chickens with little attention given to turkeys. As part of on-going surveillance for AI in southwest Nigeria, we used a competitive ELISA to detect anti-AI virus antibodies in 520 turkey sera obtained from poultry farms in Oyo, Osun, and Ondo states while haemagglutination inhibiting antibodies against low pathogenic AI viruses (LPAIVs) were detected using H3N8 and H5N2 subtype-specific antigens. The overall seroprevalence obtained by ELISA was 4.4% (23/520). Of the 23 ELISA-positive samples, 18 were positive for anti-AIV H3N8 antibodies only and four were positive for both anti-AIV H3N8 and H5N2 antibodies indicating a mixed infection, while five were negative for antibodies to either of the two AIV subtypes. Considering that turkeys have been implicated as a mixing vessel for generating influenza virus reassortants of human and avian origin, the detection of antibodies to LPAIV H3N8 and H5N2 in these turkeys is of public health concern. We advocate further studies to determine the potential role of turkeys in the zoonotic transmission of AIVs in Nigeria. Additionally, the practice of rearing turkeys with chickens should be discouraged.

1. INTRODUCTION

Avian influenza (AI) is a highly contagious disease caused by type A influenza viruses, a member of the family Orthomyxoviridae [1]. Influenza A viruses have antigenically related nucleocapsid and matrix proteins but are classified into subtypes on the basis of their haemagglutinin (H) and neuraminidase (N) antigens. Although Fouchier et al. [2] reported the presence of 16 H and 9 N subtypes of influenza A viruses, recent studies [3, 4] have identified additional H17N10 and H18N11 subtypes in bats. Many species of domestic and wild birds worldwide have been shown to be susceptible to infection with avian influenza viruses (AIVs), with aquatic birds constituting a major reservoir of these viruses which have particularly been reported to occur in poultry in either the highly pathogenic or low pathogenic forms; the overwhelming majority of isolates are of low pathogenicity for chickens and turkeys [5, 6]. Specifically, turkeys are susceptible to a wide range of influenza A viruses and, as a major exception to the host range restriction rule, are routinely infected with swine-like influenza viruses [7]. Infections in turkeys range from asymptomatic to severe disease including respiratory tract disease, depression, drop in egg production, and high mortality [7, 8].

The segmented nature of the influenza virus genome allows for reassortment of genes when a susceptible host is coinfected with different influenza virus subtypes [9], which may be from different species. This interspecies transmission of influenza viruses has been reported by several authors [9–12]. Influenza viruses with novel combinations of gene segments from different influenza-susceptible species have been isolated from turkeys [12, 13], which have been indicated to be more susceptible to AI infections than chickens [14]. Although H3- and H5-subtype influenza viruses are known to infect avian and mammalian species, including humans [15,16], the H3-subtype viruses are usually not the subject of conventional surveillance, which is biased towards detection of highly pathogenic avian strains, therefore leading to fewer H3 virus isolations from poultry flocks.

Since 2005, AI has spread from Southeast Asia to over 60 different countries, resulting in the direct death or slaughter of over 250 million poultry [17]. Recently, an outbreak of highly pathogenic avian influenza (HPAI) H5N2 that affected mostly turkey flocks was reported in several counties in Minnesota, USA [18, 19]. In Nigeria which has a poultry industry of about 160 million birds estimated at US$ 250 million [20], AI has been reported in several domestic and wild birds such as chickens [21], ducks [22], waterfowls [23], and spur-winged geese or whistling ducks [24]. Although these studies emphasize the importance of surveillance for AIV infections in the natural hosts, there has been little or no report of surveillance for the disease in turkeys, which form almost 2% of the total poultry population, in Nigeria [25].

According to the OIE [6], in serologic surveillance programs, the test to detect the anti-nucleoprotein antibody is the method commonly used because it detects antibodies to a cross-reactive antigen shared by all influenza A viruses. Some authors [26, 27] have suggested that competitive enzyme-linked immunosorbent assay (cELISA) is effective for large-scale surveillance of AIV in avian flocks or herds of other species and this test has been used to detect antibodies against different AIV strains, including H3N8 and H5N2, in chickens, ducks, turkeys, and other avian species [27]. These authors reported that since the cELISA is high in sensitivity but relatively low in specificity in turkeys, quails, and pheasants, its use for surveillance purposes should be followed by a conventional standard test (e.g., haemagglutination inhibition assay) when specific species need to be tested [27, 28]. Therefore, as part of on-going surveillance for AI in poultry, we carried out a serologic survey to investigate the prevalence of avian H3- and H5-subtype influenza virus antibodies in turkey flocks in three states of southwest Nigeria.

2. MATERIALS AND METHODS

2.1. Study Area and Survey Animals

A total of 520 (203 males and 317 females) apparently healthy turkeys sampled from 22 different flocks located in Oyo, Osun, and Ondo states, southwest Nigeria (Figure 1), were used for this study. While 315 samples were collected from 13 flocks in Oyo state, 95 were collected from six flocks in Osun state and 110 samples were collected from only three flocks in Ondo state. Additionally, the turkeys comprised 331 local and 189 exotic varieties. The southwest region plays a leading role in poultry production since an estimated 65% of Nigeria's commercial poultry population is concentrated there [25, 29].

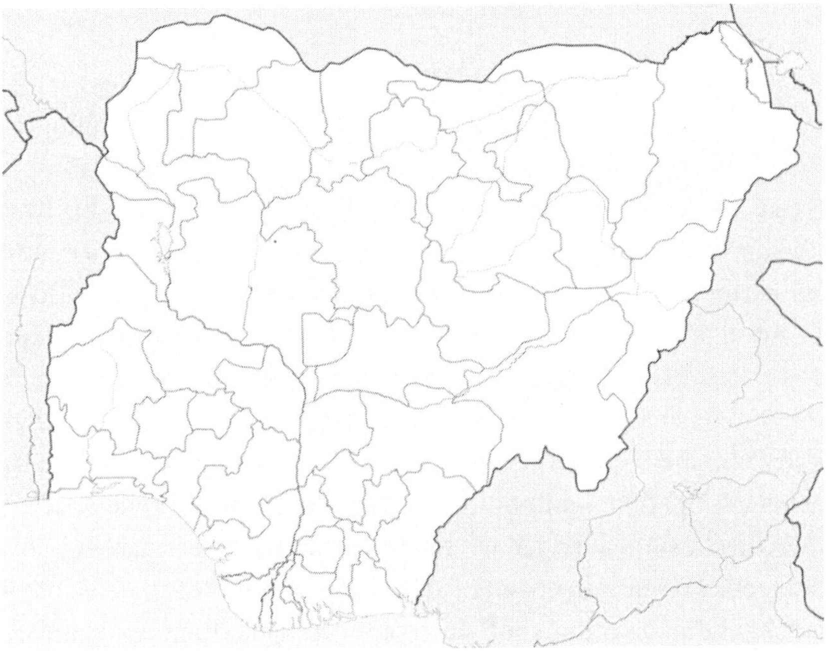

Figure 1. Map of Nigeria showing the study area (A = Oyo; B = Osun; C = Ondo).

2.2. Sample Collection

About 2 mL of blood was aseptically collected from the brachial vein of each turkey using sterile syringes and needles. The blood was allowed to clot at room temperature while sera were separated and stored at −20°C until tested. The farmers were interviewed on issues such as AI vaccination of their flocks and type of poultry they preferred to rear while the practices they employed in rearing the turkeys were observed. Additionally, the sex and breed of the turkeys as well as management system (free-range, semi-intensive, or intensive) were recorded. The turkeys were grouped based on age into growers (0–18 weeks) and adults (>18 weeks).

2.3. Serology

A competitive ELISA kit (BioNote Inc., Korea) for the quantitative detection of anti-nucleoprotein antibodies to AIV was used to screen the sera. According to the manufacturer, this kit was validated using antisera against H1–H15 subtypes of AIV. The test was performed following the kit protocol and results were read at 450 nm using a microplate ELISA reader (Optic Ivymen System, Model 2100C, Biotech SL, Madrid, Spain). The percentage inhibition (PI) for each sample was calculated from the absorbance values obtained. Positive samples (PI value ≥ 85) were screened by haemagglutination inhibition (HI) test for AIV subtype-specific antibodies using a panel of reference antigens comprising low pathogenic avian influenza (LPAI) H3N8 and H5N2 viruses and 4 haemagglutinating units of each antigen according to standard protocol [6]. Data obtained were analysed with column statistics and Fisher's exact test (two-tailed) using Graph Pad prism version 5.0 (Graph Pad software, San Diego, CA, USA) and values < 0.05 were considered significant.

3. RESULTS AND DISCUSSION

Interviews conducted with the farmers revealed that the sampled turkeys were not vaccinated against AI and rearing of turkeys with chickens was a common practice on some of the farms (Figure 2). It was also observed in two farms (one each in Oyo and Osun states) that the turkeys were kept in close proximity to pig pens while the farmers interviewed in Ondo state indicated their preference for rearing chickens instead of turkeys. Based on the ELISA, prevalence of anti-AIV antibodies in the tested turkey sera was 6.0% (19/315), 4.2% (4/95), and 0% (0/110) for Oyo, Osun, and Ondo states, respectively, with overall seroprevalence of 4.4%

(23/520). Compared to the intensively raised turkeys, those reared on free-range system had significantly higher AIV antibody prevalence (Table 1), with value of 0.034 and odds ratio (OR) of 9.4 (95% CI: 1.7–51). However, there was no significant difference in seropositivity based on state, breed, age, and sex of the birds although a greater proportion of the HI test-positive birds were intensively reared adult, female local turkeys from Oyo state (Table 2). The HI antibody titres ranged from 1 : 8 to 1 : 2048 and 1 : 64 to 1 : 2048 for LPAIV H3N8 and H5N2 subtypes, respectively. Of the tested sera, 18 were positive for anti-AIV H3N8 antibodies only and four were positive for both anti-AIV H3N8 and H5N2 antibodies, while five did not contain antibodies to either of the two AIV subtypes. Mean HI antibody titres of $\log_2$ (95% CI: 4.1–6.8) and $\log_2$ (95% CI: 5.8–13.7) were obtained for the H3N8- and H5N2-positive sera, respectively.

The current global influenza situation is characterized by a number of trends that must be closely monitored. These include an increase in the variety of animal influenza viruses cocirculating and exchanging genetic material, giving rise to novel strains [30]. Specifically, avian influenza viruses continue to be a problem worldwide because they are potentially highly infectious and can rapidly spread and cause disease in domestic poultry, and some may also infect other animal hosts, including humans [31]. Moreover, Clark and Hall [14] noted that the first sign of LPAI infection in domestic poultry is often seroconversion, which may be the only evidence of infection with some LPAI subtypes (i.e., no clinical signs present). These observations, coupled with previous reports of AIV infections in Nigeria [21–24, 32, 33], underscore the need for continuous surveillance for these infections in Nigerian poultry.

Table 1. Prevalence of avian influenza virus antibodies in sampled turkeys.

	Number sampled	Number positive (%)
State		
Oyo	315	19 (6.0)
Osun	95	4 (4.2)
Ondo	110	0 (0)
Management system		
Free-range	7	2* (28.6)
Intensive	513	21* (4.1)
Breed		
Exotic	189	10 (5.3)
Local	331	13 (3.9)
Age		
Grower	188	5 (2.7)
Adult	332	18 (5.4)
Sex		
Male	203	8 (3.9)
Female	317	15 (4.7)

*Statistically significant.
p value = 0.034, OR = 9.4 (95% CI: 1.7–51.0).

The detection of antibodies to LPAIV H3N8 and H5N2 in apparently healthy turkeys in this study indicates that LPAI H3N8 and H5N2 virus strains presently circulate in exotic and local turkey flocks in Oyo and Osun states, southwest Nigeria. Since vaccination against AI is not currently officially permitted in Nigeria and all the farmers interviewed in this study claimed that they did not vaccinate their birds against avian influenza, the antibodies detected in these birds could only have resulted from seroconversion following natural infection with the viruses. Thus, the birds could serve as reservoirs shedding the viruses into the environment, thereby playing a crucial role in the epidemiology of the disease. This finding is consistent with previous reports of infection with LPAIV H3 and H5 subtypes in poultry elsewhere [27, 34, 35] and corroborates the observation of Clark and Hall [14] that the first sign of LPAI infection in domestic poultry is often seroconversion, which may be the only evidence of infection. The nondetection of anti-AIV antibodies in turkey sera from Ondo state could be due to the fact that the samples were collected from only three flocks where the farmers interviewed indicated their preference for rearing chickens instead of turkeys.

Importantly, the detection of antibodies to both H3N8 and H5N2 LPAI viruses in some turkeys in this study is of public health concern because coinfection with different influenza viruses might provide the opportunity for reassortment, leading to the emergence of novel reassortant strains with zoonotic potential [30]. This finding is similar to that of Song et al. [27] who also detected antibodies against H3N8 and H5N2 AIV strains in chickens, turkeys and other avian species in Korea. Two of the farms (one each in Oyo and Osun states) from which seropositive birds were detected in this study also kept pigs in pens close to the turkey houses. According to Kapczynski et al. [36], coproduction of swine and turkeys on the same farm may increase the opportunity for reassortment between different AI and mammalian influenza viruses due to close contact between these species. Additionally, it has been reported [37] that if a field virus is a low pathogenic type, there is the risk of it mutating to a highly pathogenic form after circulating in susceptible poultry. Thus, the practice of rearing different animal species in the same pen or in close proximity, as observed in this study, may play a vital role in interspecies transmission of AIVs in Nigeria. It is interesting to note that five of the ELISA-positive sera did not contain antibodies to either H3N8 or H5N2 LPAI virus. This suggests the possibility of other AIV subtypes circulating among turkeys in the study area. There is therefore a need to further investigate AIV strains present in turkey flocks in southwest Nigeria using a larger panel of reference AIV subtypes as well as virus isolation and molecular identification techniques such as reverse transcriptase-polymerase chain reaction.

The detection of higher level of seropositivity to LPAI H3 virus in this study compared to the H5 subtype is significant because, unlike aquatic birds, poultry are not natural hosts of H3-subtype influenza viruses and, generally, H3-infected poultry are asymptomatic [16]. Furthermore, although the global focus remains on H5, H7, and H9 viruses as emerging disease threats for avian and mammalian species, the genetic and pathogenic changes found among H3 viruses [16] suggest a need to determine the current status of H3 AIV infections in poultry, especially among chickens and turkeys, which constitute the largest poultry species in Nigeria. Moreover, it is known that normal asymptomatic infection of avian species can silently maintain and transmit H3 influenza viruses provided there is an opportunity for genetic reassortment with other prevalent strains (e.g., H5 or H9) in avian populations [16], and this may be promoted where different avian species are kept together. Thus, mixed farming of turkeys and chickens, observed to be practiced on some farms in this study, is a critical management risk that should be discouraged. Since turkeys, like swine, have been implicated as another mixing vessel for generating influenza reassortants of human and avian origin in the field [36], additional studies are advocated to determine their potential role in the zoonotic transmission of AIV strains in Nigeria.

Table 2. ELISA-positive sera and their HI antibody titres.

Sample number	State	Management system	Breed	Age	Sex	H3N8 HI titre	H5N2 HI titre
1	Oyo	Intensive	Exotic	Adult	Male	64	64
2	Oyo	Intensive	Local	Adult	Female	64	2048
3	Oyo	Free-range	Local	adult	Female	16	2048
4	Oyo	Free-range	Local	Adult	Female	8	0
5	Oyo	Intensive	Exotic	Adult	Male	16	2048
6	Oyo	Intensive	Exotic	Adult	Female	32	0
7	Oyo	Intensive	Local	Grower	Female	16	0
8	Oyo	Intensive	Exotic	Adult	Male	16	0
9	Oyo	Intensive	Exotic	Adult	Male	0	0
10	Oyo	Intensive	Exotic	Adult	Female	0	0
11	Oyo	Intensive	Exotic	Adult	Female	8	0
12	Osun	Intensive	Local	Grower	Female	32	0
13	Osun	Intensive	Local	Adult	Male	0	0
14	Osun	Intensive	Local	Grower	Female	0	0
15	Osun	Intensive	Local	Adult	Male	0	0
16	Oyo	Intensive	Local	Adult	Male	32	0
17	Oyo	Intensive	Local	Adult	Male	8	0
18	Oyo	Intensive	Exotic	Adult	Female	8	0
19	Oyo	Intensive	Exotic	Adult	Female	32	0
20	Oyo	Intensive	Local	Adult	Female	32	0
21	Oyo	Intensive	Local	Adult	Female	2048	0
22	Oyo	Intensive	Local	Grower	Female	2048	0
23	Oyo	Intensive	Local	Adult	Female	2048	0

Figure 2. Practice of rearing turkeys with chickens.

In this study, it was observed that the odds of detecting anti-AIV antibodies were 9.4 times higher in free-range than in intensively managed turkeys. This could be due to the fact that birds (and other animals) on free-range are often neglected and allowed to scavenge for food, thus exposing them to infectious agents. According to Alexander [5], most evidence obtained on the prevalence of influenza in different types of poultry and from different geographical locations supports the view that the primary introduction is from feral birds. Therefore, influenza viruses are most likely to infect poultry reared in a way that allows contact with feral birds, such as on free-range. Based on this, we recommend that free-range rearing of turkeys in the study area be discouraged as it has the potential to favour spread of AI infections between poultry populations. Additionally, in order to better understand the epidemiology of AI in southwest Nigeria, we propose the fact that future surveillance for the disease should be done in combination with innovative tools such as social network analysis and poultry market chain analysis, both of which have successfully been used elsewhere [38–41], for the control of within-country and transboundary livestock disease outbreaks. These innovations, which should involve stakeholders like the commercial and small-holder poultry farmers, poultry sellers and butchers, and live-bird market operators, have the benefits of being able to provide a network-based approach that offers new insights into disease transmission dynamics as well as an analytical framework that allows characterization of an entire poultry industry and interlinkages among various actors in the industry [38, 42, 43]. Ultimately, this approach will serve as a basis for developing more effective strategies aimed at mitigating the risk of contracting and transmitting AI among poultry farms in Nigeria, thus reducing the socioeconomic impact of the disease on the poultry value chain.

4. CONCLUSIONS

The findings of this study not only reveal that LPAI H3N8 and H5N2 viruses presently circulate in exotic and local turkey flocks in Oyo and Osun states, southwest Nigeria, but also emphasize the importance of routine surveillance for AIVs in different avian species as part of an early warning plan for the prevention of AI outbreaks in Nigeria. We advocate that farmers be educated on the dangers inherent in the practice of free-range rearing of birds as well as raising of different avian species together or in close proximity with other animals. This knowledge will help eliminate farm practices that could otherwise have enhanced the spread or interspecies transmission of AIVs in the country. Lastly, the use of social network and market chain analyses for future AI surveillance investigations is proposed as a basis for developing more effective AI control strategies.

CONFLICT OF INTERESTS

The authors declare that there is no conflict of interests regarding the publication of this paper.

ACKNOWLEDGMENTS

This work was partly funded through the World Bank support for the Avian Influenza Control Project in Nigeria. The authors are grateful to the farmers for giving them access to their farms for sample collection.

REFERENCES

1. P. F. Wright, G. Neumann, and Y. Kawaoka, "Orthomyxoviridae," in Fields Virology, D. M. Knipe and P. M. Howley, Eds., pp. 1691–1740, Lippincott Williams & Wilkins, Philadelphia, Pa, USA, 5th edition, 2007.

2. R. A. M. Fouchier, V. Munster, A. Wallensten et al., "Characterization of a novel influenza A virus hemagglutinin subtype (H16) obtained from black-headed gulls," Journal of Virology, vol. 79, no. 5, pp. 2814–2822, 2005.

3. S. Tong, Y. Li, P. Rivailler et al., “A distinct lineage of influenza A virus from bats,” Proceedings of the National Academy of Sciences of the United States of America, vol. 109, no. 11, pp. 4269–4274, 2012.

4. S. Tong, X. Zhu, Y. Li, et al., “New world bats harbor diverse influenza A viruses,” PLoS Pathogens, vol. 9, no. 10, Article ID e1003657, 2013.

5. D. J. Alexander, “A review of avian influenza in different bird species,” Veterinary Microbiology, vol. 74, no. 1-2, pp. 3–13, 2000.

6. Office International des Epizooties, “Avian influenza,” in OIE Manual of Diagnostic Tests and Vaccines for Terrestrial Animals, chapter 2.3.4, OIE, Paris, France, 2012.

7. D. E. Swayne and D. J. King, “Avian influenza and Newcastle disease,” Journal of the American Veterinary Medical Association, vol. 222, no. 11, pp. 1534–1540, 2003.

8. D. L. Suarez, P. R. Woolcock, A. J. Bermudez, and D. A. Senne, “Isolation from turkey breeder hens of a reassortant H1N2 influenza virus with swine, human, and avian lineage genes,” Avian Diseases, vol. 46, no. 1, pp. 111–121, 2002.

9. T. Kuiken, E. C. Holmes, J. McCauley, G. F. Rimmelzwaan, C. S. Williams, and B. T. Grenfell, “Host species barriers to influenza virus infections,” Science, vol. 312, no. 5772, pp. 394–397, 2006.

10. C. W. Olsen, S. Carey, L. Hinshaw, and A. I. Karasin, “Virologic and serologic surveillance for human, swine, and avian influenza virus infections among pigs in the north-central United States,” Archives of Virology, vol. 145, no. 7, pp. 1399–1419, 2000.

11. D. L. Suarez, E. Spackman, and D. A. Senne, "Update on molecular epidemiology of H1, H5, and H7 influenza virus infections in poultry in North America," Avian Diseases, vol. 47, pp. 888–897, 2003.

12. Y. K. Choi, J. H. Lee, G. Erickson et al., "H3N2 influenza virus transmission from swine to turkeys, United States," Emerging Infectious Diseases, vol. 10, no. 12, pp. 2156–2160, 2004.

13. Y. Tang, C. W. Lee, Y. Zhang et al., "Isolation and characterization of H3N2 influenza A virus from turkeys," Avian Diseases, vol. 49, no. 2, pp. 207–213, 2005.

14. L. Clark and J. Hall, "Avian influenza in wild birds: status as reservoirs, and risks to humans and agriculture," Ornithological Monographs, vol. 60, pp. 3–29, 2006.

15. K. Van Reeth, "Avian and swine influenza viruses: our current understanding of the zoonotic risk,"Veterinary Research, vol. 38, no. 2, pp. 243–260, 2007.

16. J. Pu, G.-Z. Zhang, J.-H. Ma et al., "Serologic evidence of prevalent avian H3 subtype influenza virus infection in chickens," Avian Diseases, vol. 53, no. 2, pp. 198–204, 2009.

17. Food and Agriculture Organization (FAO), "Questions and answers: The facts of bird flu," 2007

18. Animal and Plant Health Inspection Service (APHIS), "USDA confirms highly pathogenic H5N2 avian influenza in Stearns County, Minnesota," 2015, http://content.govdelivery.com/accounts/USDAAPHIS/bulletins/fb549d.

19. Animal and Plant Health Inspection Service (APHIS), "USDA confirms more highly pathogenic H5N2 avian influenza in 13

flocks in Minnesota," 2015, http://content.govdelivery.com/accounts/USDAAPHIS/bulletins/1008951.

20. Federal Department of Livestock and Pest Control Services, "Current issues in the control of avian influenza in Nigeria," in Proceedings of the Stakeholders Meeting with Poultry Association, Presented by M. D. Saidu, Abuja, Nigeria, October 2007.

21. C. O. Aiki-Raji, A. I. Adebiyi, V. I. Agbajelola, C. O. Aiki-Raji, et al., "Surveillance for low pathogenic avian influenza viruses in live-bird markets in Oyo and Ogun States, Nigeria," Asian Pacific Journal of Tropical Diseases, vol. 5, no. 5, pp. 369–373, 2015.

22. T. Coker, C. Meseko, G. Odaibo, and D. Olaleye, "Circulation of the low pathogenic avian influenza subtype H5N2 virus in ducks at a live bird market in Ibadan, Nigeria," Infectious Diseases of Poverty, vol. 3, no. 1, article 38, 2014.

23. C. A. Meseko, A. T. Oladokun, and B. Shehu, "An outbreak of Highly Pathogenic Avian Influenza (HPAI) in a mixed farm by the introduction of a water fowl," Nigerian Veterinary Journal, vol. 28, no. 3, pp. 67–69, 2007.

24. C. J. Snoeck, A. T. Adeyanju, S. De Landtsheer et al., "Reassortant low-pathogenic avian influenza H5N2 viruses in African wild birds," Journal of General Virology, vol. 92, no. 5, pp. 1172–1183, 2011.

25. D. F. Adene and A. E. Oguntade, Poultry Sector Country Review, FAO Animal Production and Health Division, 2006.

26. E. Starick, O. Werner, H. Schirrmeier, B. Köllner, R. Riebe, and E. Mundt, "Establishment of a competitive ELISA (cELISA) system for the detection of influenza A virus

nucleoprotein antibodies and its application to field sera from different species," Journal of Veterinary Medicine Series B: Infectious Diseases and Veterinary Public Health, vol. 53, no. 8, pp. 370–375, 2006.

27. D. S. Song, Y. J. Lee, O. M. Jeong et al., "Evaluation of a competitive ELISA for antibody detection against avian influenza virus," Journal of Veterinary Science, vol. 10, no. 4, pp. 323–329, 2009.

28. M. Petric, L. Comanor, and C. A. Petti, "Role of the laboratory in diagnosis of influenza during seasonal epidemics and potential pandemics," Journal of Infectious Diseases, vol. 194, supplement 2, pp. S98–S110, 2006.

29. Federal Republic of Nigeria, "Avian influenza control and human pandemic preparedness and response project," National Baseline Survey Final Report, 2007

30. World Health Organization, Influenza, World Health Organization (WHO), Geneva, Switzerland, 2015,http://www.who.int/influenza /publications/warningsignals201502/en/.

31. A. C. Morales Jr., D. A. Hilt, S. M. Williams et al., "Biologic characterization of H4, H6, and H9 type low pathogenicity avian influenza viruses from wild birds in chickens and turkeys," Avian Diseases, vol. 53, no. 4, pp. 552–562, 2009.

32. D. F. Adene, A. M. Wakawa, P. A. Abdu, et al., "Clinico-pathological and husbandry features associated with the maiden diagnosis of avian influenza in Nigeria," Nigerian Veterinary Journal, vol. 27, no. 1, pp. 32–38, 2006.

33. T. M. Joannis, C. A. Meseko, A. T. Oladokun et al., "Serologic and virologic surveillance of avian influenza in Nigeria, 2006-7," European Surveillance, vol. 13, no. 42, 2008.

34. C.-W. Lee, D. A. Senne, and D. L. Suarez, "Effect of vaccine use in the evolution of Mexican lineage H5N2 avian influenza virus," Journal of Virology, vol. 78, no. 15, pp. 8372–8381, 2004.

35. L. Stipkovits, L. Egyed, V. Palfi et al., "Effect of low-pathogenicity influenza virus H3N8 infection onMycoplasma gallisepticum infection of chickens," Avian Pathology, vol. 41, no. 1, pp. 51–57, 2012.

36. D. R. Kapczynski, E. Gonder, K. Liljebjelke, R. Lippert, D. Petkov, and B. Tilley, "Vaccine-induced protection from egg production losses in commercial turkey breeder hens following experimental challenge with a triple-reassortant H3N2 Avian influenza virus," Avian Diseases, vol. 53, no. 1, pp. 7–15, 2009.

37. O. Werner and T. C. Harder, "Avian influenza," in Influenza Report, B. S. Kamps, C. Hoffman, and W. Preiser, Eds., chapter 2, pp. 48–86, Flying Publisher, Paris, France, 2006.

38. J. Humphrey and L. Napier, "The value chain approach as a tool for assessing distributional impact of standards on livestock markets: guidelines for planning a programme and designing case studies," in FAO AGA/ESC Initiative on Market Exclusion, Institute of Development Studies, Brighton, UK, 2005.

39. V. Martin, X. Zhou, E. Marshall et al., "Risk-based surveillance for avian influenza control along poultry market chains in South China: the value of social network analysis," Preventive Veterinary Medicine, vol. 102, no. 3, pp. 196–205, 2011.

40. G. Fournié, J. Guitian, S. Desvaux et al., "Interventions for avian influenza A (H5N1) risk management in live bird market

networks," Proceedings of the National Academy of Sciences of the United States of America, vol. 110, no. 22, pp. 9177–9182, 2013.

41. C. Moore, G. S. Cumming, J. Slingsby, and J. Grewar, "Tracking socioeconomic vulnerability using network analysis: insights from an avian influenza outbreak in an ostrich production network," PLoS ONE, vol. 9, no. 1, Article ID e86973, 2014.

42. A. S. Klovdahl, "Social networks and the spread of infectious diseases: the AIDS example," Social Science and Medicine, vol. 21, no. 11, pp. 1203–1216, 1985.

43. R. Kaplinsky, "Spreading the gains from globalization: what can be learned from value-chain analysis?"Problems of Economic Transition, vol. 47, no. 2, pp. 74–115, 2004.

CHAPTER 9

Phosphorus Bioavailability: A Key Aspect for Conserving this Critical Animal Feed Resource with Reference to Broiler Nutrition

Xiuhua Li *, Dagong Zhang, Tsung Y. Yang † and Wayne L. Bryden

Poultry Science Unit, School of Agriculture and Food Sciences, The University of Queensland, Gatton Qld 4343, Australia

ABSTRACT

Phosphorus (P) is an essential element, and the majority of animal feed phosphate is derived from phosphate rock that is a non-renewable resource. Current global P reserves may be depleted in 50–100 years. This poses the challenge of securing future P supply for the global animal feed industries. Currently, nutritionists formulate diets with substantial safety margins to guarantee that animals do not become P deficient. Excessive dietary P concentrations increase, not only the cost of diets, but also P excretion and pollution of the environment. We contend that understanding P bioavailability is central to the sustainable use of this mineral in animal agriculture. Poultry accounts for approximately 50% of animal feed phosphate consumption worldwide and for this reason we use the meat chicken or broiler as a case study to explore the nuances of P bioavailability. We conclude that, to tackle the challenge of dietary P bioavailability, cooperative research on a global scale is needed to standardise measurement procedures in order to produce a robust and reliable database which can be used by nutritionists to formulate diets to meet the bird's P requirements precisely. Achievement of this goal will assist endeavours to sustain the global supply of phosphorus.

KEYWORDS

phosphorus; requirements; available P; bioavailability; poultry; animal nutrition; rock phosphate; global feed supply

1. INTRODUCTION

Phosphorus (P) is an essential element for all forms of life, from single to multicellular organisms. It is required for normal muscle growth and egg formation, is an important component of nucleic acids, the genetic code, and phospholipids, and is also a co-factor of many enzyme systems. Phosphorus plays a vital role in maintaining osmotic and acid-base balance, energy metabolism, amino acid metabolism and protein synthesis. Phosphorus was named "life's bottleneck" by chemist and science writer Isaac Asimov [1], who claims: "Life can multiply until all the phosphorus has gone, and then there is an inexorable halt which nothing can prevent," also writing, "We may be able to substitute nuclear power for coal, and plastic for wood, and yeast for meat, and friendliness for isolation—but for P there is neither substitute nor replacement."

The main ingredients of non-ruminant diets are cereal grains and their by-products. These diets usually require supplementation with inorganic P, because of the low concentrations, and low availability of total P in cereal feed ingredients. Meat and bone meal has relatively high bioavailable P content but is banned from use in animal feed in European countries. This policy increases the demand for inorganic feed phosphates [2]. The majority of feed phosphate is derived from phosphate rock that is a non-renewable resource and becoming increasingly scarce and expensive [3]. Current global P reserves may be depleted in 50–100 years [4,5]. This poses the challenge of securing future P supply for the international and national animal feed industries. Poultry account for approximately 50% of animal feed phosphate consumption worldwide [6]. Feed phosphates make a significant contribution to the P content of any broiler diet; generally providing as much as 60% of the bird's P requirements [7]. There is limited information on biologically determined P values in feed ingredients for poultry [8,9,10]. Therefore, nutritionists have to formulate diets with substantial safety margins to guarantee that birds do not become P deficient. Greater dietary P concentrations increase the cost of poultry production, and P excretion into manure. Excessive amounts of P in manure contribute to the pollution of lakes, streams and wetlands leading to surface water eutrophication. Eutrophication of fresh water globally is accelerated by P accumulation [11,12].

All these concerns have attracted much global attention and stimulated the re-examination of the use of inorganic P in animal diets. This issue is pertinent to all non-ruminant animals that require dietary P supplementation, especially poultry (broilers, layers and breeders) and pigs. In this review, the focus is on meat chickens or broilers as much of the P used in animal nutrition is in poultry diets. In an effort to optimise P use by poultry, the P requirements of the modern broiler have been re-examined [13,14,15,16,17,18] and defining P bioavailability in feedstuffs has commenced [2,19]. Mention of bioavailability, which encompasses digestion, absorption and utilisation of P, evokes much debate and is the thrust of this review.

In many instances, the terms "availability" and "bioavailability" are interchangeable when referring to nutrients. However, to avoid confusion in this review, we use the term bioavailability to encompass the digestion, absorption and utilisation of P, as the term "available P" has a specific connotation in P nutrition as described below. We contend that bioavailability is central to the sustainable use of P in animal agriculture. A thorough understanding of P bioavailability will permit the development of strategies to enhance efficiency of P utilisation, reduce wastage of P resources and reduce environmental pollution by P.

2. PHOSPHORUS METABOLISM

It is important when discussing phosphorus requirements and bioavailability to appreciate the metabolism of phosphorus (Figure 1) and to remember the close association that phosphorus homeostasis has with the metabolism of two other nutrients; calcium (Ca) and vitamin D. The major features of these interactions are briefly reviewed here and the interested reader is referred to detailed reviews on the topic [20,21]. The major store of Ca and P is the skeleton where the cationic and anionic forms, respectively, of these minerals connect to form hydroxyapatite, which confers rigidity on the bone matrix [2]. Bone is continually turned over and to ensure that the biological demands for both Ca and P are met, metabolism of these essential minerals is tightly coordinated.

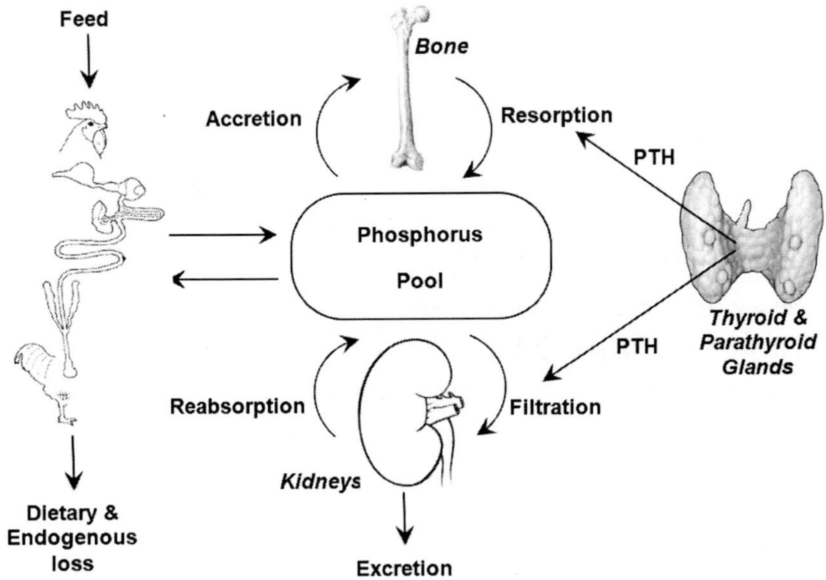

Figure 1. Phosphorus metabolism. Coordination of three organ systems, intestine, kidneys and bone, is central to Ca and P homeostasis. Reduced circulating concentrations of Ca causes increased secretion of parathyroid hormone (PTH) by the parathyroid glands. PTH increases bone resorption and renal Ca reabsorption but increased renal excretion of P. In the kidneys, PTH also stimulates formation of the hormonal form of vitamin D_3 which increases intestinal absorption of Ca and, to a lesser extent, the absorption of P. Elevated blood Ca concentrations as a result of these processes creates a negative feedback loop to ensure homeostasis. If hypercalcaemia occurs, calcitonin is secreted by the thyroid gland to inhibit bone resorption.

Utilisation of Ca and P is modulated by the relative amounts of each in the diet [2]. However, the quantities of each available for their respective metabolic functions is determined by efficiencies of intestinal absorption, glomerular filtration, renal tubular reabsorption, rates of transfer from blood to bone, and intestinal endogenous losses [20]. The efficiencies of these processes in the different tissues are modulated by several hormones, chiefly parathyroid hormone (PTH) and the hormonal form of vitamin D_3 (1,25 dihydroxycholecalciferol; calcitriol) [20,21]. The control of P metabolism differs from that of Ca, with less control in the gut and greater control in the kidneys. Moreover, there is increasing evidence that dietary protein and carbohydrate affect intestinal metabolism of Ca and P [21]. Together, the factors that determine intestinal absorption and kidney excretion along with endocrine regulation of absorption and reabsorption in the kidneys facilitate P homeostasis.

3. PHOSPHORUS REQUIREMENTS

Research on the P requirement of broilers has been the subject of numerous investigations for many decades; however, the minimum requirement for this nutrient has still not been established with any certainty. The published results are variable as different breeds of birds, age of birds, feed ingredients (with or without phytase supplementation), Ca and P sources were used in different studies. In some instances, the differences observed may reflect true differences in requirements, not merely variation resulting from experimental conditions. To further complicate interpretation of results, different criteria, including body weight gain, feed conversion ratio (FCR), tibia breaking strength, tibia ash, or toe ash contents have been used to assess responses to P (Table 1).

Table 1. Summary of available P requirements (g/kg diet) of broilers.

Criteria of Assessment	AP (g/k) (Bird Age, Day) 1–21	21–42	Reference
Carcass P analysis	3.7 (day 0–10), 3.0 (day 10–30)	2.4 (day 30–40)	[2]
Bone breaking strength	3.9	3.0	[13]
FCR		1.63	[22]
FCR	3.3–3.9		[23]
Growth and tibia ash	3.2–2.8 (day 18–32)	2.4–1.9 (day 32–42), 1.1 (day 43–49)	[24,25]
Growth, bone ash and processing losses	4.5	4.0	[26]
Weight gain	3.5 (day 0–14)	3.0 (day 15–49)	[15,18]
Weight gain	3.2–3.4		[27]
Weight gain		1.86	[22]
Weight gain	3.2–3.5		[23]
Weight gain	2.8–2.9		[28]
Tibia or toe ash	4.5	3.5	[29]
Tibia ash		3.3	[22]
Tibia ash	3.5–3.9		[23]
Tibia ash	3.7–3.9	3.1 (day 49)	[30]

Evidence in the literature clearly demonstrates that the P requirement of broilers is much lower than NRC [29] recommendations and the values currently used by the industry [2,7,13,15,17,18,22,23,24,25,27,28,31]. The greatest challenge in applying a lower dietary P strategy in practice is to decide what dietary concentrations of P to use in a formulation. This is due to a lack of information on biologically determined P values in feed ingredients for poultry and an agreed P evaluation system [8,9,10].

4. EVALUATION OF PROCEDURES THAT DETERMINE BIOAVAILABLE PHOSPHORUS

There are many approaches for determining P bioavailability of feed and feed ingredients. Some methods are biological and rely on *in vivo* measures including growth, bone strength, bone ash content and blood parameters. Chemical procedures have also been developed to estimate P bioavailability. Most published values using both biological and chemical procedures are qualitative rather than quantitative. Recently, there has been an international move to establish ileal digestible P as the method of choice for determining P bioavailability. An overview of the different approaches for the determination P bioavailability is given below.

4.1. Biologically Determined Values

Initial attempts to quantitate the availability of P for chicks from various phosphate sources used tibia bone ash as the response factor. The method was based on relative percentage of tibia bone ash of chicks fed a test supplement to that of chicks fed a reference standard, thus obtaining relative bioavailability P values [32]. A positive linear relationship was found between tibia bone ash percentage and dietary P intake. A similar relationship was found with toe ash percentages [33,34,35]. Hoshii and Yoshida [36] demonstrated that P bioavailability determined with a slope ratio assay using toe ash was closely associated with the values determined by carcass P retention, suggesting that changes in toe ash concentration reflect changes in net P utilisation.

In the 1980s and '90s, the bioavailability of P was routinely estimated by non-linear bioassays (asymptotic or sigmoidal curves) of body weight and tibia bone or toe ash, using a graded range of P supplements [37,38,39,40,41]. Others [42,43] determined P bioavailability using responses in bone ash content, bone breaking strength, body weight gain or blood inorganic P concentration of birds fed a test P source compared with the response of birds to a standard reference P source. Bioassays based on growth, blood parameters, tibia bone or toe ash percentage and tibia bone breaking strength can only provide a relative biological reference value or ranking [42,43] and do not measure true or apparent P bioavailability *per se*. The values do not define the quantity of bioavailable P and therefore have limited value when formulating diets [44].

Early attempts to determine true or apparent P retention by collecting excreta in balance studies with chicks were unsuccessful [45,46]. Later, Simons and Versteegh [47] were successful using a balance assay based on excreta collection, when birds were fed a basal diet, nearly free of P (5 g/kg Ca), to which test phosphates were added at low concentrations of approximately 2 g/kg AP. At this level, there is negligible urinary P excretion [47].

4.2. Chemically Determined Values

The most common P evaluation system (NRC, [29,48]) is based on the relationship of P with phytate; non-phytate P (NPP) and phytate P (PP). Phytate is the bound form in which P is found in plants; it is discussed in detail elsewhere (see [49,50,51,52,53]). The NPP system has the advantage of being based on the chemical analysis of feed ingredients. Non-phytate P is the P in the feed or ingredient that is not bound as part of phytic acid. It can be determined by subtracting analysed PP from analysed total P [14]. With this approach, it is assumed that NPP, together with P from non-plant sources are completely available to poultry and non-ruminant animals; assuming that PP is not available to non-ruminant animals. If PP in plant feedstuffs is not analysed, it is assumed that NPP is 30% of total plant P and the remaining 70% is PP. Non-phytate P has been referred to as "AP" or used interchangeably with "AP." These assumptions based on simple differentiation from chemical analysis have been adopted in diet formulation worldwide, since publication by the NRC in 1977.

The NPP system has limitations. Chemical analysis determines how much P is present, but does not indicate to what degree, if any, P is utilised by animals and birds. In other words, chemical analysis does not provide information on P bioavailability of the ingredient or diet and thus any changes in PP and P bioavailability of an ingredient will not be reflected in this system. Numerous studies have found that NPP is not 100% bioavailable and up to 80% of PP can be bioavailable to birds [8,54,55]. In fact, no elements are completely absorbed and utilised by poultry [56].

Phytate has been found in the excreta of the chicks fed a phytate-free (casein-gelatin) diet, whereas none was found in the excreta of germ-free chicks fed the same diet. It appears that phytate may be synthesised by the microflora in the avian digestive tract [57]. Therefore, it is possible that the net microbial hydrolysis of phytate occurring during transit of digesta through gastrointestinal tract may be partly offset by microbial synthesis [58]. This makes determination of dietary NPP requirement by total collection or indigestible marker and analysing NPP in excreta unreliable since some NPP may have been incorporated into phytate. Other chemical or *in vitro*approaches have attempted to relate the solubility of P supplements to P bioavailability determined *in vivo* and have been unsuccessful [59].

4.3. Ileal Digestibility as an Estimate of Phosphorus Bioavailability

In an attempt to overcome the difficulties associated with current approaches to P bioavailability determinations, the Nutrition Working Group of the European Federation of Branches of World's Poultry Science Association (hereafter Rodehutscord and WPSA, [60]) has developed a protocol to determine ileal P digestibility by regression analysis. Two to three diets are formulated for each test P source. A low P basal diet is used, and a minimum of two levels of the P source under test are added to the test diet. At the highest level of inclusion, P supply must not exceed the requirement. With such conditions, the AP of the P source under test can be determined from the slopes of linear regression equations (see [60] for details). The protocol is currently being evaluated in an international collaborative ring test. Collinearity appears to exist between two variables used in the regression analysis and may lead to unreliable and unstable estimates of regression coefficients [61].

Finally, none of the methods outlined above measure P bioavailability *per se* but rather provide an estimate. Accurate measurements of bioavailability would require very expensive and laborious experiments, including the use of radioactive isotopes.

What is required is a robust procedure that will provide reliable and repeatable estimates of P bioavailability. It would appear that precaecal P digestibility is the method of choice. A similar situation exists with amino acid availability where precaecal or ileal digestibility is used to estimate availability. Detailed discussions that describe amino acid digestibility assays are available [62].

5. TERMS USED TO DESCRIBE BIOAVAILABLE PHOSPHORUS

Over the years, a number of terms have been suggested to define certain aspects of P utilisation as described by Rodehutscord and WPSA [60] and outlined below. These terms have largely arisen to reflect differing approaches to the determination of bioavailable P.

Digestible P is the portion of dietary total P that is not recovered in faeces. The determination of digestible P requires *in vivo* studies with quantitative determination of dietary P intake and P excretion in faeces. Alternatively, digestible P can be determined by using an indigestible marker. However, as bird excreta consists of faeces and urine, determination of digestible P requires colostomised birds to exclude urinary P content.

Retainable P (RP) indicates the proportion of dietary total P that is retained in the body and can be determined quantitatively following measurement of P intake and P excretion in faeces [60]. It is also called total tract P digestibility and is easy to measure. However, as urine is included in the determination of digestible P, this term is not an appropriate description of the measurement in poultry. Importantly, if a P intake is lower than the requirement, the P content of the urine can be ignored. In these circumstances, the determination of RP should be similar to the determination of digestible P.

Available P (AP) is the quantity of P that is absorbed from the diet by the animal. This definition of AP is different from previous interpretations [29,60] where it has been equated to NPP and RP. In the currently acceptable definition of AP [60], it is determined by including an indigestible marker in the feed and is the proportion of total dietary P that is not recovered in the terminal ileum. It is also called precaecal digestible P or ileal P digestibility [60]. The major advantage of this method is that if P intake exceeds requirements, urinary excretion of P does not confound the results.

There is now general agreement that ileal digestible P is the preferred approach for determining AP as a quantitative estimate of P bioavailability. Rodehutscord and WPSA [60] recommend that retainable P values not be used in the future to avoid confusion. Retainable P reflects both P digestibility and post-absorptive utilisation since bird excreta includes faeces and urine. Dietary supply of P relative to requirement and body status affects total tract P utilisation [63]. In several studies of individual feed ingredients and mixed diets, the RP in total tract is substantially different from AP in the ileum (Table 2). Based on these and other publications that demonstrate a similar trend, both utilisation of P or Ca in feeds and requirement in birds should be expressed on an ileal digestible basis [63]. Furthermore, AP can be used to measure the marginal level of P supplied in animal diets to meet the P requirement of the animal [60].

Table 2. Comparisons of retainable P (RP) with available P (AP) in the ileum of poultry [63].

Ingredient	RP (%)	AP (%)	Reference
Low-phosphorus mixed diet	43	29	[64] [1]
Low-phytate soybean meal	77	94	[65] [2]
Conventional soybean meal	60	94	[65] [2]
Low-phosphorus mixed diet	59	52	[66] [1]
Low-phosphorus mixed diet	45	30	[67] [1]
Canola meal	39	66	[63] [2]

1 Expressed as apparent digestibility

2 Expressed as true digestibility derived from regression analysis.

6. PHOSPHORUS SOURCES AND AVAILABILITY

Sources of P in poultry diets include feed ingredients of both plant and animal origin and also inorganic forms of the mineral. These sources differ in P bioavailability.

6.1. Plant Feedstuffs

Plant feedstuffs include cereal grains, oil seeds and the respective by-products. Phytate is the principal storage form of P in many plant tissues. As commercial broiler diets are primarily comprised of feedstuffs of plant origin, most of the P present in these ingredients is in the form of phytate P (PP) [68]. The concentration of PP within the same ingredient can vary considerably. Barrier-Guillot *et al.* [69] found that the concentration of PP ranged from

0.092% to 0.268% dry matter (DM) in wheat and depended on soil fertility, plant variety, and stage of plant maturity at harvest.

The bioavailability of PP for broilers ranges from 0% to 80% (Table 3) [2] depending on diets, bird age and metabolic adaptation. The degree to which PP is utilised by birds will depend, to a large extent, upon the hydrolysis of PP in the gastrointestinal tract. Moreover, the bioavailability of PP in different feed ingredients ranges from 0% to 60% [8,47,67], while the bioavailability of total P within ingredients varies considerably (15% to 75%; Li *et al.*, unpublished data [70]). There are many factors affecting PP utilisation: phytase activity of dietary ingredients, dietary concentrations of fibre, Ca, P and vitamin D_3, age of bird and genotype [58,71].

6.2. Animal Feedstuffs

There are a number of animal-sourced feedstuffs (fish meals, meat meals, meat and bone meals) used in poultry diets mainly as protein supplements. Among these, meat meal is also a potential P source depending on its bone content. Meat meals are prepared by rendering offal from abattoirs. There can be a wide variation between batches depending on the raw materials used and rendering conditions [72]. If the ash content is high, this indicates that it contains a higher amount of bone and is referred to as meat and bone meal. If the ash content is low it is referred to as meat meal.

Meat and bone meal is a variable source of protein for poultry diets, but can be an excellent source of Ca and P and other minerals. Therefore, it is commonly used as a P supplement in poultry diets [7]. However, there are few reports that have measured AP from animal-derived feedstuffs. Limited RP values of animal-derived feedstuffs are shown in Table 4.

Table 3. Total P (TP) (g/kg), retainable P (RP) (% of TP), non-phytate P (NPP) (g/kg), phytate P (PP) (g/kg), PP (% of TP) and calculated PP degradability (%) of some of the feedstuffs in three-week-old broilers [2].

Feedstuffs	TP (g/kg)	RP (% TP) *	NPP (g/kg)	PP (g/kg)	PP (% TP)	PP Degradability (%) **
Beans	4.9	52	1.3	3.6	73.5	53
Lupin	3	72	1.5	1.5	50	80
Maize	3	29	0.7	2.3	76.7	16
Peas	4.1	41	1.5	2.6	63.4	23
Rapeseed	10.9	33	3.8	7.1	65.1	10
Rice bran	17.2	16	3.1	14.1	82	2
Soy bean (heat treated)	5.5	54	2	3.5	63.6	49
Soybean meal	7.1	61	2.8	4.3	60.6	61
Sunflower meal	11.9	38	4.2	7.7	64.7	19
Wheat	3.4	48	0.9	2.5	73.5	46
Wheat middlings	10.8	36	2.8	8	74.1	26

* Values were determined in broilers fed diets containing 1.8 g calculated retainable P and 5.0 g Ca per kg diet from 10 days of age onward. P retention was measured in a three-day balance period (from days 21 to 24) in which dietary P intake and excreta P output were measured quantitatively; ** Phytate P degradability was calculated as: 100 × (AP/0.8–NPP)/PP [2].

Table 4. Total P (TP) (g/kg), retainable P (RP) (% of TP) of animal-derived feedstuffs in three-week-old broilers [2].

Animal Feedstuffs	TP (g/kg)	RP (% TP) *
Bone meal	76	59
Fish meal	22	74
Meat meal	29	65
Meat and bone meal	60	66

* Values were determined in broilers fed diets containing 1.8 g calculated retainable P and 5.0 g Ca per kg diet from 10 days of age onward. P retention was measured in a three-day balance period (from days 21 to 24) in which dietary P intake and excreta P output were measured quantitatively.

6.3. Inorganic Phosphorus Sources

Inorganic P sources are commonly derived from natural rock phosphates. These natural phosphates are not readily utilised by poultry or other animals but must be chemically modified before inclusion in animal feed. Rock phosphates are changed into various orthophosphate forms: monocalcium phosphate, dicalcium phosphate, monodicalcium phosphate, defluorinated rock phosphate and monosodium phosphate [44]. Some commonly used mineral sources and their respective P composition and retention rate are listed in Table 5.

Most of these studies determined apparent retention of P from each of the inorganic sources within the deficiency range of P. Leske and Coon [13] showed dramatic reductions in P retention from monocalcium phosphate as dietary P concentration approached the requirement level (98% retention at half of the P requirement to

59% retention at requirement level). Waldroup [73] reported that nearly 50% of excreted P is likely to be of inorganic origin.

Table 5. Inorganic feed phosphates and their respective total P (TP) (g/kg) and retainable P (RP) (% of TP) [2].

Inorganic P Source	TP (g/kg)	RP (% TP) *	Reference
Calcium sodium phosphate	180	59	[2] *
Dicalcium phosphate	183	83	[74] [1,2]
Dicalcium phosphate (anhydrous)	197	55	[2] *
Dicalcium phosphate (hydrous)	181	77	[2] *
Monocalcium phosphate	226	84	[2] *
Monodicalcium phosphate (hydrous)	213	79	[2] *
Monodicalcium phosphate	203	77	[13] [1]
Monodicalcium phosphate	200	80	[13] [1]
Monodicalcium phosphate	216	81	[13] [1]
Monosodium phosphate	224	92	[2] *
Defluorinated phosphate	182	86	[74] [1,2]
Defluorinated phosphate	179	76	[74] [1,2]

* Values were determined in broilers fed diets containing 1.8 g calculated retainable P and 5.0 g Ca per kg diet from 10 days of age onward. P retention was measured in a three-day balance period (from days 21 to 24) in which dietary P intake and excreta P output were measured quantitatively; 1 Apparent utilisation of P from inorganic sources by broiler as determined under deficiency conditions; 1,2 Retainable P determined through broken line slope response.

Phosphorus must be in the phosphate form to be absorbed by birds. Improper processing and higher temperatures have a detrimental impact on the biological availability of calcium phosphates. As phosphates are heated, pyro- and meta- complexes are formed which greatly reduce the P bioavailability of inorganic sources. The availability of inorganic P sources is substantially affected by hydration, with hydrated phosphate sources generally more biologically available than anhydrous sources of the same type [75]. Feed phosphate with larger particle size has been shown to be more biologically available. Larger particle phosphate is retained longer in the gizzard allowing more complete digestion [76].

7. FACTORS AFFECTING PHOSPHORUS AVAILABILITY

Determination of the bioavailability of P in feedstuffs is a major challenge. There are many factors affecting P bioavailability in birds, including: experimental techniques; the chemical form of P; dietary concentrations of Ca, vitamin D3, energy, protein, fat, and vitamin K; the availability and interaction of other nutrients in the diet, feed processing, and particle size; bird physiological and health factors, such as feed consumption, growth rate, sex and age; management factors, including ambient temperature and lighting program. Some of these factors have been investigated in detail [14] and the major factors are briefly described below.

7.1. Phosphorus and Phytate

High levels of dietary inorganic P reduce PP availability [49,77] due to either the inhibitory effect of P on phytase activity [78], a negative feedback mechanism by excessive amounts of dietary P, or inhibition by products of phytate hydrolysis [49,58].

Phytate is the principal storage form of P in plant feedstuffs and is considered to be an anti-nutritive factor [79]. Phytate carries a strong negative charge and is capable of binding di- and trivalent cations such as calcium, cobalt, copper, iron, magnesium, manganese, nickel and zinc in very stable complexes [58,80], reducing the availability of P as well as these minerals to the animal [2,54,55,81].

Phytate has been shown to inhibit activities of some digestive enzymes such as pepsin, α-amylase and trypsin. Phytate may inhibit proteolysis by changing the protein configuration of digestive enzymes [82]. Inhibition may also result from the chelation of Ca ions which are essential co-factors for the activity of trypsin and α-amylase leading to reduced protein and starch digestion [50,83,84,85,86].

It has also been reported that phytate can reduce fat digestibility by forming insoluble Ca-phytate complexes with fatty acids in the gut lumen [87]. In its chelated form, the phytate molecule is difficult to hydrolyse by phytase. The pH affects the solubility of phytate. Most phytate mineral complexes are soluble at low pHs (less than 3.5) with maximum insolubility occurring between 4 and 7 [88]. Champagne [89] found that Ca-phytate complexes precipitate at pH between 4 and 6, which is the approximate pH of the intestine where the Ca ions should be absorbed. Taylor [90] has suggested that the primary factor affecting PP utilisation is the Ca ion concentration in the small intestine where insoluble Ca-phytate complexes form. A precipitated phytate mineral complex would not be accessible for hydrolysis or absorption in the intestine.

7.2. Calcium

Calcium is an essential element for bone and egg shell formation, blood clotting, muscle contraction and transmission of nerve impulses. Calcium is also an important co-factor for many enzymes and hormones.

Increases in dietary P and Ca concentrations may affect apparent digestion of these nutrients. High levels of dietary Ca and low levels of P had detrimental effects on broiler performance [16,17,91,92]. Elevated dietary Ca levels increase pH in the gut and, as a result, P absorption [93] and retention [91] are decreased. High dietary Ca concentration can significantly increase gastrointestinal pH [94] and decrease not only phytate hydrolysis, but also pepsin activity in the proventriculus/gizzard and reduce apparent ileal protein digestibility [51]. High plasma P levels decrease Ca absorption from the gut [95]. The optimum dietary Ca-to-P ratio is of utmost importance for broiler production. With the Ca: P ratio, there will be a range of values that can meet metabolic demands for any production outcome. When dietary P concentration is high, birds can tolerate a wider range of Ca levels than when dietary P marginally meets the requirement (Li*et al.* unpublished data [96]).

Total Ca values are currently used for feed formulation since the bioavailability of Ca for feed ingredients have not been determined. Bioavailable Ca from plant raw materials is expected to be low. This is due to the high phytate content of these raw materials and the low contribution of Ca to the diet. Also, to some extent, phytate will negatively influence the availability of Ca from this dietary source by binding Ca in phytate-mineral complexes. Nevertheless, there is a need to determine the bioavailability of Ca in feed ingredients to further delineate the interaction of this nutrient with other nutrients and to improve the precision of feed formulation.

7.3. Vitamin D

Vitamin D_3 is essential for absorption of Ca and P utilization as depicted in Figure 1. Several reports have demonstrated that dietary addition of vitamin D_3 can significantly enhance the retention of P in birds [52]. However, excess vitamin D_3 and its metabolites have not been shown to have a further beneficial effect on bird productivity when birds are already consuming adequate vitamin D_3. Like many other minerals, understanding of the molecular mechanisms of P absorption, especially the role of the hormonal form of vitamin D_3, is still limited.

7.4. Phytase

Numerous studies have demonstrated that dietary addition of phytase increases hydrolysis of PP and improves P utilisation in pigs and poultry [15,16,17,18,53]. Maximum utilisation of phytate by broilers fed diets with supplemental phytase was approximately 50% on average [97] following dietary P supplementation. There are many factors affecting the efficacy of microbial phytase such as phytase source and dose [98], physical factors of feedstuffs, feed particle size, Ca to P ratio [99], Ca bioavailability and inorganic P supplements, vitamin D status [100], bird age [101] and body weight [102], gut pH, digesta retention time, and Ca and P status. Phytase was found to be most efficacious when added to diets containing low AP and adequate Ca concentrations (Li *et al.*unpublished data [96]).

Dietary Ca level affects the efficacy of phytase in broilers [103], but not to the same extent for all phytases [104]. The same source of phytase can produce different responses in different flocks [105,106]. Moreover, the efficacy of feed phytase may be improved by the simultaneous dietary addition of other exogenous enzymes. All these variables make predicting phytase responses very difficult and lead to variation in P availability.

7.5. Birds

The ability of poultry to absorb dietary P is known to be influenced by age and physiological state. Young chicks with a rapidly developing skeletal system are critically sensitive to P deficiency [107] and tend to use P more efficiently than do mature birds [108]. Older birds have a greater ability to use PP, possibly due to more phytase activity in their intestinal tract [56]. A concern often raised by commercial nutritionists is the relevance of digestibility values generated with birds of one age group to birds of another age or other species.

Applegate *et al.* [106] conducted two experiments in which chicks were fed the same diet and found that the apparent PP hydrolysis varied nearly two-fold between the first and second experiment. Similar results have been reported by Nelson and Kirby [109] and Tamim *et al.* [103]. Bird variations make it difficult to predict treatment responses of different flocks. The variation between experiments and between flocks arises from many sources, some of which are summarised in Table 6.

Table 6. Summary of variation in processes associated with feeding of phosphorus to broilers [10].

Source of Variation	Coefficient of Variation (%)
Sampling variation	5 to 10
Analytical variation	5
Mixer variation	5 to 10
Bird utilisation	16
Ingredient variation (corn and soybean meal)	8 to 13

8. CONCLUSIONS

Re-examination of the phosphorus requirement of modern broilers has demonstrated that it is lower than the NRC [29] recommendations and the values currently used by the chicken meat industry. However, there are practical limitations in applying this information to industry as there is limited data on the bioavailability and variability of P in feed ingredients; there are few reports that have measured P bioavailability of feedstuffs for poultry. Furthermore, the different methods used have made comparisons between the studies and application to practice extremely difficult. It is only when commercial poultry nutritionists have this information that they will be in a position to reduce the safety margins for P when formulating diets. To tackle these challenges, cooperative research on a global scale is needed to standardise measurement procedures in order to produce a robust and reliable database which can be used by nutritionists to formulate diets to meet the bird's P requirement precisely; this approach is appropriate as only a couple broiler strains dominate international markets. As a starting point, the universal adoption of ileal P digestibility [60] as the preferred procedure to determine AP would be a major step forward. Moreover, to make sense of the data derived and to allow comparison of the results, it will be necessary to report ingredient composition of experimental diets, dietary concentrations of Ca and non-phytate P, dietary electrolyte balance and mineral content, all of which influence P absorption. Achievement of these goals will assist endeavours to sustain the global supply of phosphorus.

ACKNOWLEDGMENTS

The Australian Government Rural Industries Research and Development Corporation -Chicken Meat Program, Canberra, ACT, has supported the phosphorus research conducted by the authors.

AUTHOR CONTRIBUTIONS

All the authors contributed equally to the paper.

CONFLICTS OF INTEREST

The authors declare no conflicts of interest.

REFERENCES

1. Asimov, I. *Asimov on Chemistry*; Doubleday: Garden City, NY, USA, 1974.

2. Van der Klis, J.D.; Versteegh, H.A.J. Phosphorus nutrition of poultry. In *Recent Developments in Poultry Nutrition 2*; Garnsworthy, P.C., Wiseman, J., Eds.; Nottingham University Press: Nottingham, UK, 1999; pp. 309–320.

3. Shastak, Y.; Witzig, M.; Hartung, K.; Rodehutscord, M. Comparison of retention and prececal digestibility

measurements in evaluating mineral phosphorus sources in broilers. *Poult. Sci.***2012**, *91*, 2201–2209.

4. Cordel, D.; Drangert, J.; White, S. The story of phosphorus: Global food security and food for thought. *Glob. Environ. Chang.* **2009**, *19*, 292–305.

5. Neset, T.S.; Cordell, D. Global phosphorus scarcity: Identifying synergies for a sustainable future. *J. Sci. Food Agric.* **2012**, *92*, 2–6.

6. Devereux, C.; Smart, M.; Kalt, F.P.; Takei, N. Animal feeds: Phosphate supplements. In *CEH Marketing Research Report*; HIS Chemical: Englewood, CO, USA, 1994.

7. Waldroup, P.W. Nutritional approaches to reducing phosphorus excretion by poultry. *Poult. Sci.* **1999**, *78*, 683–691.

8. Van der Klis, J.D.; Versteegh, H.A.J. Phosphorus nutrition in broilers. In *Recent Advances in Animal Nutrition*; Garnsworty, P.C., Wiseman, J., Haresign, W., Eds.; Nottingham University press: Nottingham, UK, 1996; pp. 71–83.

9. Cossa, J.; Oloffs, K.; Kluge, H.; Jeroch, H. Investigation into the TP and PP content in different varieties of grain maize. In Proceedings of the 11th European Symposium on Poultry Nutrition; World's Poultry Science Association: Faaberg, Denmark, 1997; pp. 444–446.

10. Applegate, T.J.; Angel, R. Feasibility *versus* Practicality of Phosphorus Reduction in Poultry: Progress and Future Needs. USDA National Agricultural Library: West Lafayette, IN, USA. Available online: http://www.cals.ncsu.edu/waste_ mgt/natlcenter/sanantonio/ApplegateandAngel.pdf (accessed on 8 December 2014).

11. Schindler, D.W. Evolution of phosphorus limitation in lakes. *Sciences* **1977**, *195*, 260–262.

12. Sharpley, A.N.; Chapra, S.C.; Wedepohl, R.; Sims, J.T.; Daniel, T.C.; Reddy, K.R. Managing agricultural phosphorus for protection of surface waters: Issues and options. *J. Environ. Qual.* **1994**, *23*, 437–451.

13. Leske, K.; Coon, C.N. The development of feedstuff retainable phosphorus values for broilers. *Poult. Sci.* **2002**, *81*, 1681–1693.

14. Angel, R. Calcium and phosphorus requirements in poultry. In Proceedings of the 1st International Phytase Summit, Washington, DC, USA, 28–30 September 2010; pp. 65–71.

15. Li, X.; Zhang, D.; Huang, K.H.; Bryden, W.L. Available phosphorus requirement of broilers fed sorghum based diet from day 22 to 49. In Proceedings of the 19th European Symposium on Poultry Nutrition, Potsdam, Germany, 26–29 August 2013; p. 200.

16. Li, X.; Zhang, D.; Huang, K.H.; Bryden, W.L. Dietary calcium levels and non-phytate phosphorus requirement of broilers from days 1 to 21. *Aust. Poult. Sci. Symp.* **2013**, *24*, 50.

17. Yang, T.Y.; Setiyawan, E.; Li, X.; Zhang, D.; Bryden, W.L. Dietary calcium levels and non-phytate phosphorus requirement of broilers from days 22–49. In Proceedings of Recent Advances in Animal Nutrition in Australia 2013; University of New England: Armidale, New South Wales, Australia, 2013.

18. Zhang, D.; Li, X.; Huang, K.H.; Bryden, W.L. Available phosphorus requirement of broilers fed sorghum based diet from day 1 to 21. In Proceedings of the 19th European

Symposium on Poultry Nutrition, Potsdam, Germany, 26–29 August 2013; p. 196.

19. Li, X.; Zhang, D.; Bryden, W.L. Bioavailability of phosphorus in feedstuffs for broilers. In*Final Report for Rural Industries Research and Development Corporation*; Rural Industries Research and Development Corporation: Kingston, Australia, 2015.

20. Klasing, K.C. *Comparative Avian Nutrition*; CAB International: Wallingford, UK, 1998.

21. Vitti, D.M.S.S., Kebreab, E., Eds.; *Phosphorus and Calcium Utilization and Requirements in Farm Animals*; CAB International: Wallingford, UK, 2010.

22. Yan, F.; Kersey, J.H.; Waldroup, P.W. Phosphorus requirements of broiler chicks three to six weeks of age as influenced by phytase supplementation. *Poult. Sci.* **2001**, *80*, 455–459.

23. Persia, M.E.; Saylor, W.W. Effects of broiler strain, dietary nonphytate phosphorus, and phytase supplementation on chick performance and tibia ash. *J. Appl. Poult. Res.* **2006**, *15*, 72–81.

24. Angel, R.; Applegate, T.J.; Christman, M. Effect of dietary non-phytate phosphorus (nPP) on performance and bone measurements in broilers fed on a four-phase feeding system. *Poult. Sci.* **2000**, *79* (Suppl. 1), 21–22.

25. Angel, R.; Applegate, T.J.; Christman, M.; Mitchell, A.D. Effect of dietary non-phytate phosphorus (nPP) level on broiler performance and bone measurements in the starter and grower phase. *Poult. Sci.* **2000**, *79* (Suppl. 1), 22.

26. Moran, E.T., Jr.; Todd, M.C. Continuous submarginal phosphorus with broilers and the effect of preslaughter transportation: Carcass defects, further-processing yields, and tibia-femur integrity. *Poult. Sci.* **1994**, *73*, 1448–1457.

27. Waldroup, P.W.; Kersey, J.H.; Saleh, E.A.; Fritts, C.A.; Yan, F.; Stilborn, H.L.; Crum, R.C., Jr.; Raboy, V. Non-phytate phosphorus requirement and phosphorus excretion of broilers fed diets composed of normal or high available phosphate corn with and without microbial phytase. *Poult. Sci.* **2000**, *79*, 1451–1459.

28. Yan, F.; Waldroup, P.W. Non-phytate phosphorus requirement and phosphorus excretion of broiler chicks fed diets composed normal or high available phosphate corn as influenced by phytase supplementation and vitamin D source. *Int. J. Poult. Sci.* **2006**, *5*, 219–228.

29. NRC. *Nutrient Requirements of Poultry*, 9th ed.; National Academy Press: Washington, DC, USA, 1994.

30. Yan, F.; Kersey, J.H.; Fritts, C.A.; Waldroup, P.W. Phosphorus requirements of broiler chicks six to nine weeks of age as influenced by phytase supplementation. *Poult. Sci.* **2003**, *83*, 294–300.

31. Dhandu, A.S.; Angel, R. Broiler non-phytin phosphorus requirement in the finisher and withdrawal phases of a commercial four-phase feeding system. *Poult. Sci.* **2003**, *82*, 1257–1265.

32. Gillis, M.B.; Norris, L.C.; Heuser, G.F. Studies on the biological value of inorganic phosphates. *J. Nutr.* **1954**, *52*, 115–125.

33. Grau, C.R.; Zweigart, P.A. Phosphatic clay as a phosphorus source for chicks. *Poult. Sci.* **1953**, *32*, 500–503.

34. Fritz, J.C.; Roberts, T.; Bochne, J.W.; Hove, E.L. Factors affecting the chick's requirement for phosphorus. *Poult. Sci.* **1969**, *48*, 307–320.

35. Yoshida, M.; Hoshii, H. Improvement of biological assay to determine available phosphorus with growing chicks. *J. Poult. Sci.* **1977**, *14*, 33–43.

36. Hoshii, H.; Yoshida, M. Comparison of phosphorus availabilities estimated biologically based on toe ash content and carcass phosphorus retention. *J. Poult. Sci.* **1977**, *14*, 279–283.

37. Potchanacom, M.; Potter, L.M. Biological values of phosphorus from various sources for young turkeys. *Poult. Sci.* **1987**, *66*, 505–513.

38. Potter, L.M. Bio-availability of phosphorus from various phosphates based on body weight and toe ash measurements. *Poult. Sci.* **1988**, *67*, 96–102.

39. Potter, L.M.; Potchanacom, M.; Ravindran, V.; Kornegay, E.T. Bio-availability of phosphorus in various phosphate sources using body weight and toe ash as response criteria. *Poult. Sci.* **1995**, *74*, 813–820.

40. Ketels, E.; De Groote, G. The relative bio-availability and the ileal digestibility of phosphorus in mineral and animal sources. In Proceedings of the 18th World's Poultry Congress, Nagoya, Japan, 4–9 September 1988; pp. 873–875.

41. De Groote, G.; Ketels, P.; Seynaeve, P. Relative bio-availability of phosphorus in meat meals as influenced by

particle size. In Proceedings of the International Poultry Congress, Istanbul, Turkey, 6–10 September 1991; pp. 42–52.

42. De Groote, G. Biological availability of phosphorus in feed phosphates for broilers. In Proceedings of the 4th European Symposium on Poultry Nutrition, Tours, France, 17–20 October 1983; Larbier, M., Ed.; pp. 91–102.

43. Sauveur, B. Bio-availability to poultry of plant-origin phosphorus-Methodological criticisms and results. In Proceedings of the 4th European Symposium on Poultry Nutrition, Tours, France, 17–20 October 1983; Larbier, M., Ed.; pp. 103–113.

44. Payne, S.G. The Phosphorus Availability of Feed Phosphates in Broilers. MS Thesis, University of Stellenbosh, Stellenbosh, South Africa, 2005.

45. Edwards, H.M., Jr.; Gillis, M.B. A chromic oxide balance method for determining phosphate availability. *Poult. Sci.* **1959**, *38*, 569–574.

46. Sibbald, I.R. Measurement of mineral bio-availability: Extension of true metabolisable energy methodology. *Poult. Sci.* **1982**, *61*, 485–487.

47. Simons, P.C.M.; Versteegh, H.A.J.; Jongbloed, A.W.; Kemme, P.A.; Slump, P. Improvement of phosphorus availability by microbial phytase in broilers and pigs. *Br. J. Nutr.* **1990**, *64*, 525–540.

48. NRC. *Nutrient Requirements of Poultry*, 7th ed.; National Academy Press: Washington, DC, USA, 1977.

49. Ankra-Badu, G.A.; Aggrey, S.E.; Pesti, G.M.; Bakalli, R.I.; Edwards, H.M., Jr. Modelling of parameters affecting phytate

phosphorus bioavailability in growing birds. *Poult. Sci.* **2004**,*83*, 1083–1088.

50. Thompson, L.U.; Yoon, H. Starch digestibility as affected by poly phenols and phytic acid. *J. Food Sci.* **1984**, *49*, 1228–1229.

51. Walk, C.L.; Bedford, M.R.; McElroy, A.P. Influence of limestone and phytase on broiler performance, gastrointestinal pH, and apparent ileal nutrient digestibility. *Poult. Sci.* **2011**,*91*, 1371–1378.

52. Ravindran, V.; Bryden, W.L.; Kornegay, E.T. Phytates: Occurrence, bioavailability and implications in poultry nutrition. *Poult. Avian Biol. Rev.* **1995**, *6*, 125–143.

53. Selle, P.H.; Ravindran, V. Microbial phytase in poultry nutrition. *Anim. Feed Sci. Technol.***2007**, *135*, 1–41.

54. Angel, R.; Saylor, W.W.; Mitchell, A.D.; Powers, W.; Applegate, T.J. Effect of dietary phosphorus, phytase, and 25-hydroxycholecalciferol on broiler chicken bone mineralization, litter phosphorus, and processing yields. *Poult. Sci.* **2006**, *85*, 1200–1211.

55. Rodehutscord, M.; Dieckmann, A.; Witzig, M.; Shastak, Y. A note on sampling digesta from the ileum of broilers in phosphorus digestibility studies. *Poult. Sci.* **2012**, *91*, 965–971.

56. Axe, D. Phosphorus value in ingredient sources examined. *Feedstuffs* **1998**, *70*, 17–28.

57. Savage, J.E.; Yohe, J.M.; Pickett, E.E.; O'Dell, B.L. Zinc metabolism in the growing chick. Tissue concentration and effect of phytate on absorption. *Poult. Sci.* **1964**, *43*, 420–426.

58. Wise, A. Dietary factors determining the biological activities of phytase. *Nutr. Abs. Rev.* **1983**,*53*, 791–806.

59. Shastak, Y.; Rodehtscord, M. Determination and estimation of phosphorus availability in growing poultry and their historical development. *World's Poult. Sci. J.* **2013**, *69*, 569–586.

60. Rodehutscord, M.; WPSA (Working Group No. 2: Nutrition of the European Federation of Branches of World Poultry Science Association). Determination of phosphorus availability in poultry. *World's Poult. Sci. J.* **2013**, *69*, 687–698.

61. Allison, P. When Can You Safely Ignore Multicollinearity? 2012. Available online: http://statisticalhorizons.com/ multicollinearity (accessed on 4 May 2015).

62. Bryden, W.L.; Li, X. Amino acid digestibility and poultry feed formulation: Expression, limitations and application. *Rev. Bras. Zootec.* **2010**, *39* (Suppl.), 279–287.

63. Adeola, O.; Applegate, T.J. Phosphorus and calcium equivalency of phytase. In Proceedings of the 1st International Phytase Summit, Washington, DC, USA, 28–30 September 2010; pp. 45–64.

64. Adedokun, S.A.; Sands, J.S.; Adeola, O. Determining the equivalent phosphorus released by an *Escherichia coli* derived phytase in broiler chicks. *Can. J. Anim. Sci.* **2004**, *84*, 437–444.

65. Dilger, R.N.; Adeola, O. Estimation of true phosphorus digestibility and endogenous phosphorus loss in growing chicks fed conventional and low-phytate soybean meals. *Poult. Sci.* **2006**, *85*, 661–668.

66. Jendza, J.A.; Dilger, R.N.; Sands, J.S.; Adeola, O. Efficacy and equivalency of an Escherichia coli-derived phytase for

replacing inorganic phosphorus in the diets of broiler chickens and young pigs. *J. Anim. Sci.* **2006**, *84*, 3364–3374.

67. Nyannor, E.K.D.; Adeola, O. Corn expressing an *Escherichia coli*-derived phytase gene: Comparative evaluation study in broiler chicks. *Poult. Sci.* **2008**, *87*, 2015–2022

68. Nelson, T.S. The hydrolysis of phytate phosphorus by chicks and laying hens. *Poult. Sci.***1976**, *55*, 2262–2264.

69. Barrier-Guillot, B.; Casado, P.; Maupetit, P.; Jondreville, C.; Gatel, F. Wheat phosphorus availability: 1-*In vitro* study; Factors affecting endogenous phytasic activity and phytic phosphorus content. *J. Sci. Food Agric.* **1996**, *70*, 62–68.

70. Li, X.; Zhang, D.; Bryden, W.L. University of Queensland: Gatton, Australia Unpublished Work. , 2015. University of Queensland: Gatton, Australia Unpublished Work. , 2015.

71. De Groote, G.; Huyghebaert, G. The bio-availability of phosphorus from feed phosphates for broilers as influenced by bio-assay method, dietary Ca-level and feed form. *Anim. Feed Sci. Technol.* **1997**, *69*, 329–340.

72. Skurray, G.R. The nutritional evaluation of meat meals for poultry. *World's Poult. Sci. J.* **1974**,*30*, 129–136.

73. Waldroup, P.W. Phosphorus, phytase, and the environment—A retrospective look. *Proc. MD Nutr. Conf. Feed Manuf.* **2002**, *49*, 195–202.

74. Coon, C.N.; Seo, S.; Manangi, M.K. The determination of retainable phosphorus, relative biological availability, and relative biological value of phosphorus sources for broiler. *Poult. Sci.* **2007**, *86*, 857–868.

75. Gillis, M.B.; Edwards, H.M., Jr.; Young, R.J. Studies on the availability of calcium orthophosphates for chickens and turkeys. *J. Nutr.* **1962**, *78*, 155–162.

76. Lima, F.R.; Mendonca, C.X., Jr.; Alvarez, J.C.; Garzillo, J.M.F.; Ghion, E.; Leal, P.M. Biological evaluations of commercial dicalcium phosphates as sources of available phosphorus for broiler chicks. *Poult. Sci.* **1997**, *76*, 1707–1713.

77. Sanders, A.M.; Edwards, H.M., Jr.; Roland, G.N. Calcium and phosphorus requirements of the chick as determined response surface analysis. *Br. J. Nutr.* **1992**, *67*, 21.

78. Ravindran, V.; Cabahug, S.; Ravindran, G.; Selle, P.H.; Bryden, W.L. Response of broiler chickens to microbial phytase supplementation as influenced by dietary phytic acid and non-phytate phosphorus levels. II. Effects on apparent metabolisable energy, nutrient digestibility and nutrient retention. *Br. Poult. Sci.* **2000**, *41*, 193–200.

79. Bryden, W.L.; Selle, P.H.; Ravindran, V.; Acamovic, T. Phytate: An anti-nutrient factor in animal diets. In *Poisonous Plants: Global Research and Solutions*; Panter, K.E., Wierenga, T.L., Pfister, J.A., Eds.; CABI Publishing: Wallingford, UK, 2007; pp. 279–284.

80. Maenz, D.D.; Engle-Schaan, C.M.; Newkirk, R.W.; Classen, H.L. The effect of minerals and mineral chelators on the formation of phytase-resistant and phytase susceptible forms of phytic acid in solution and in a slurry of canola meal. *Anim. Feed Sci. Technol.* **1999**, *81*, 177–192.

81. Pallauf, J.; Rimbach, G.; Pippig, S.; Schindler, B.; Most, E. Effect of phytase supplementation to a phytate-rich diet based on wheat, barley and soya on the bioavailability of dietary

phosphorus, calcium, magnesium, zinc and protein in piglets. *Agribiol. Res.* **1994**, *47*, 39–48.

82. Singh, M.; Krikorian, A.D. Inhibition of trypsin activity *in vitro* by phytase. *J. Agric. Food Chem.* **1982**, *30*, 799–800.

83. Deshpande, S.S.; Cheryan, M. Effects of phytate, divalent cations and their implications on alpha-amylase activity. *J. Food Sci.* **1984**, *49*, 516–519.

84. Liener, I.E. Anti-nutritional factors in legume seeds: State of art. In *Recent Advances of Research on Anti-Nutritional Factors in Legume Seeds*; Huismen, J., Van Der Poel, T.F.B., Liener, I.E., Eds.; Pudoc: Wageningen, The Netherlands, 1989; pp. 6–13.

85. Knuckles, B.E.; Betschart, A.A. Effect of phytate and other myo-inositol phosphate esters on alpha-amylase digestion of starch. *J. Food Sci.* **1987**, *52*, 719–721.

86. Caldwell, R.A. Effect of calcium and phytic acid on the activation of trypsinogen and the stability of trypsin. *J. Agric. Food Chem.* **1992**, *40*, 43–46.

87. Leeson, S.; Summers, J.D.; Caston, L. Response of brown egg strain layers to dietary calcium or phosphorus. *Poult. Sci.* **1993**, *72*, 1510–1514.

88. Selle, P.H.; Ravindran, V.; Caldwell, R.A.; Bryden, W.L. Phytate and phytase: Consequences for protein utilisation. *Nutr. Res. Rev.* **2000**, *13*, 255–278.

89. Champagne, E.T. Effects of pH on mineral-phytate, protein-mineral-phytate and mineral-fiber interactions. Possible consequences of atrophic gastritis on mineral bioavailability from high fiber foods. *J. Am. Coll. Nutr.* **1988**, *38*, 199–215.

90. Taylor, T.G. Dietary phosphorus and egg shell thickness in the domestic fowl. *Br. Poult. Sci.* **1965**, *6*, 79–87.

91. Keshavarz, K. The effect of dietary levels of calcium and phosphorus on performance and retention of these nutrients by laying hens. *Poult. Sci.* **1986**, *65*, 114–121.

92. Keshavarz, K. Phosphorus requirement of laying hens with and without phytase. In Proceedings of the Cornell Nutrition Conference for Feed Manufacturers, Syracuse, NY, USA, 19–21 October 1999; pp. 81–92.

93. Hurwitz, S.; Bar, A. Absorption of calcium and phosphorus along the intestinal tract of the laying fowl as influenced by dietary calcium and eggshell formation. *J. Nutr.* **1965**, *86*, 433–438.

94. Guinotte, F.; Gautron, J.; Nys, Y.; Soumarmon, A. Calcium solubilisation and retention in the gastrointestinal tract in chicks (*Gallus domesticus*) as a function of gastric acid secretion inhibition and of calcium carbonate particle size. *Br. J. Nutr.* **1995**, *73*, 125–139.

95. Keshavarz, K.; Austic, R.E. Effects of dietary minerals on acid–base balance and eggshell quality in chickens. *J. Nutr.* **1990**, *120*, 1360–1369.

96. Li, X.; Zhang, D.; Bryden, W.L.; Huang, K.H. University of Queensland: Gatton, Australia Unpublished Work. , 2013. University of Queensland: Gatton, Australia Unpublished Work. , 2013.

97. Selle, P.H.; Ravindran, V.; Bryden, W.L.; Scott, T. Influence of dietary phytate and exogenous phytase on amino acid digestibility in poultry: A review. *J. Poult. Sci.* **2006**, *43*, 89–103.

98. Shirley, R.B.; Edwards, H.M., Jr. Graded levels of phytase past industry standards improves broiler performance. *Poult. Sci.* **2003**, *82*, 671–680.

99. Manangi, M.K.; Coon, C. Phytate phosphorus hydrolysis in broilers in response to dietary phytase, calcium, and phosphorus concentrations. *Poult. Sci.* **2008**, *87*, 1577–1586.

100. Mohammed, A.; Gibney, M.J.; Taylor, T.G. The effects of dietary levels of inorganic phosphorus, calcium and cholecalciferol on the digestion of phytate phosphorus by the chick. *Br. Poult. Sci.* **1991**, *66*, 251–259.

101. Kornegay, E.T. Nutritional, environmental and economic considerations for using phytase in pig and poultry diets. In *Nutrient Management of Food Animals to Enhance and Protect the Environment*; Kornegay, E.T., Ed.; Lewis Publishers: New York, NY, USA; CRC Press: Boca Raton, FL, USA, 1996; pp. 277–302.

102. Punna, S.; Roland, D.A., Sr. Influence of supplemental microbial phytase on first cycle laying hens fed phosphorus-deficient diets from day one of age. *Poult. Sci.* **1999**, *78*, 1407–1411.

103. Tamim, N.M.; Angel, R.; Christman, M. Influence of dietary calcium and phytase on phytate phosphorus hydrolysis in broiler chickens. *Poult. Sci.* **2004**, *83*, 1358–1367.

104. Augspurger, N.R.; Baker, D.H. Phytase improves dietary calcium utilisation in chicks, and oyster shell, carbonate, citrate, and citrate-malate forms of calcium are equally bioavailable.*Nutr. Res.* **2004**, *24*, 293–301.

105. Angel, R.; Dhandu, A.S.; Applegate, T.J. Phosphorus requirements of broilers and the impact of exogenous phytases.

In Proceedings of the Arkansas Poultry Nutrition Conference; University of Arkansas: Fayetteville, AR, USA, 2002; pp. 1–22.

106. Applegate, T.J.; Angel, R.; Classen, H.L. Effect of dietary calcium, 25-hydroxycholecalciferol, or bird strain on small intestinal phytase activity in broiler chickens. *Poult. Sci.* **2003**, *82*, 1140–1148.

107. Nelson, T.S.; Peeler, H.T. The availability of phosphates from single and combined phosphates to chicks. *Poult. Sci.* **1961**, *40*, 1321–1327.

108. Waldroup, P.W. Bioassays remain necessary to estimate phosphorus, calcium bioavailability.*Feedstuffs* **1996**, *68*, 13–23.

109. Nelson, T.S.; Kirby, L.K. The calcium binding properties of natural phytate in chick diets.*Nutr. Rep. Int.* **1987**, *35*, 949–956.

CHAPTER 10

Metabolism of Deoxynivalenol and Deepoxy-Deoxynivalenol in Broiler Chickens, Pullets, Roosters and Turkeys

Heidi E. Schwartz-Zimmermann [1,*], Philipp Fruhmann [2,3], Sven Dänicke [4], Gerlinde Wiesenberger [5], Sylvia Caha [1], Julia Weber [2] and Franz Berthiller [1]

[1]Christian Doppler Laboratory for Mycotoxin Metabolism, Center for Analytical Chemistry, Department of Agrobiotechnology (IFA-Tulln), University of Natural Resources and Life Sciences, Vienna, Tulln 3430, Austria

[2]Institute of Applied Synthetic Chemistry, Vienna University of Technology, Vienna 4060, Austria

[3]Centre of Electrochemical Surface Technology, Wiener Neustadt 2700, Austria

[4]Institute of Animal Nutrition, Friedrich-Loeffler-Institute (FLI), Federal Research Institute for Animal Health, Braunschweig 38116, Germany

[5]Department of Applied Genetics and Cell Biology, University of Natural Resources and Life Sciences, Vienna, Tulln 3430, Austria

ABSTRACT

Recently, deoxynivalenol-3-sulfate (DON-3-sulfate) was proposed as a major DON metabolite in poultry. In the present work, the first LC-MS/MS based method for determination of DON-3-sulfate, deepoxy-DON-3-sulfate (DOM-3-sulfate), DON, DOM, DON sulfonates 1, 2, 3, and DOM sulfonate 2 in excreta samples of chickens and turkeys was developed and validated. To this end, DOM-3-sulfate was chemically synthesized and characterized by NMR and LC-HR-MS/MS measurements. Application of the method to excreta and chyme samples of four feeding trials with turkeys, chickens, pullets, and roosters confirmed DON-3-sulfate as the major DON metabolite in all poultry species studied. Analogously to DON-3-sulfate, DOM-3-sulfate was formed after oral administration of DOM both in turkeys and in chickens. In addition, pullets and roosters metabolized DON into DOM-3-sulfate. *In vitro* transcription/translation assays revealed DOM-3-sulfate to be 2000 times less toxic on the ribosome than DON. Biological recoveries of DON and DOM orally administered to broiler chickens, turkeys, and pullets were 74%–106% (chickens), 51%–72% (roosters), and 131%–151% (pullets). In pullets, DON-3-sulfate concentrations increased from jejunum chyme samples to excreta samples by a factor of 60. This result, put into context with earlier studies, indicates fast and efficient absorption of DON between crop and jejunum, conversion to DON-3-sulfate in intestinal mucosa, liver, and possibly kidney, and rapid elimination into excreta via bile and urine.

KEYWORDS

mycotoxins; biomarkers; HPLC-MS/MS; poultry; deoxynivalenol-3-sulfate; deoxynivalenol sulfonates

1. INTRODUCTION

After being discovered in 1972 [1], the Fusarium mycotoxin deoxynivalenol (DON) is one of the most common contaminants of cereal-based food and feed. Since its discovery, a lot of research has been carried out on its occurrence [2,3], toxicity [4], and reduction [5,6] as well as on its metabolization by plants [7], humans [8], and animals [9]. Nevertheless, the current knowledge, particularly on metabolization by different animal species, is far from complete.

One metabolization type common to most studied animal species is de-epoxidation. Formation of deepoxy-DON (DOM) is predominantly achieved by the gut microbiota. Its extent varies greatly both between species and also between individuals of one species [9,10]. The second metabolization pathway of DON observed in most of the investigated animal species is glucuronidation. Glucuronidation is carried out by endogenous UDP-glucuronosyltransferases in liver, and possibly also in intestinal microsomes [11]. In general, the extent of glucuronidation and the regiospecificity of the reaction are species-dependent with additional great individual differences [11,12,13].

Both de-epoxidation and glucuronidation have been studied as metabolization pathways of DON for many years. In humans and pigs, de-epoxidation is of minor importance [14,15,16,17,18], whereas glucuronidation is a major metabolization step [8,14,18,19,20]. In ruminants [21,22] and in rats [23], both de-epoxidation and glucuronidation of DON and of the formed DOM are significant metabolic processes. In addition, very recently two independent research groups discovered sulfonation as the main metabolization pathway of DON in rats [24,25]. Poultry, however, shows only minor metabolization of DON by glucuronidation, de-epoxidation, and sulfonation [26]. In contrast, sulfation was recently discovered as a major metabolization pathway, with

DON-3α-sulfate as the predominant metabolic product of DON in chickens and turkeys [25,26]. According to Wan and co-workers who used radioisotope counting radio-HPLC for quantitation, DON-3-sulfate in excreta of chickens accounted for 89% ± 7% of the administered dose of DON (2.5 mg/kg body weight (b.w.)) [25]. Very recently, the toxicokinetic behavior of DON after oral application to turkeys and broiler chickens was investigated, and the peak areas of DON-3-sulfate and DON in plasma were compared [26]. Five minutes after administration of DON, the DON-3-sulfate to DON ratio in plasma was already 38 for turkeys and almost 2000 for broiler chickens. Thirty minutes after treatment, ratios increased to 141 and to nearly 10,000 for turkeys and broilers, respectively. These data indicate rapid absorption as well as very rapid and extensive metabolization of DON to DON-3-sulfate in both investigated avian species.

Although DOM is a minor metabolite in poultry [27,28], de-epoxidation of DON-3-sulfate or rapid sulfation of DOM may still be significant processes. We therefore hypothesized that DOM-3-sulfate might equally be a natural DON metabolite in poultry. The first aim of our work was therefore to synthesize and characterize DOM-3-sulfate and DOM-15-sulfate and to investigate their toxicity on ribosomes, the molecular targets of trichothecenes. The second aim was to develop and validate the first LC-MS/MS-based method for simultaneous quantitation of DON, DOM, DON-, and DOM-sulfate as well as DON- and DOM-sulfonates (DONS and DOMS) in excreta of different poultry species. The third objective of the work was to prove the hypothesis of formation of DON-3-sulfate and DOM-3-sulfate, as well as of DONS and DOMS in poultry consuming feed contaminated with DON and/or DOM. To this end, excreta and, in one occasion, chyme samples from four previous feeding trials with turkeys, chickens, pullets [28], and roosters [29,30] were re-analyzed. In three of these trials, feed consumption was recorded and the biological recoveries of orally administered DON and DOM could be assessed. In addition, the effect of species (turkey and chicken) and concomitant infection with *Ascaridia galli* on DON metabolization could be investigated. The results of this work will contribute to the understanding of DON metabolization in poultry.

2. RESULTS

2.1. Synthesis, Purification, and Characterization of DOM-3-Sulfate and DOM-15-Sulfate

Similarly to production of DON-sulfates [31], synthesis of DOM-sulfates is a two-step reaction. In the first step, three protected intermediates were produced: 2,2,2-trichloroethyl-DOM-3-sulfate (71.0 mg, 29%), 2,2,2-trichloroethyl-DOM-15-sulfate (14.0 mg, 6%) and, as by-product, bis(2,2,2-trichloroethyl) DOM-3,15-disulfate (30.0 mg, 9%). These substances were recovered as white solid with an overall yield of 44%. NMR data of the protected intermediates are given in the electronic supplementary material; NMR spectra are shown in Figure S1. Deprotection of the intermediates and column chromatography with ammonium hydroxide as a mobile phase additive yielded 36.7 mg (67% of the protected intermediate) of DOM-3-sulfate, 5.3 mg (49%) of DOM-15-sulfate (both as ammonium salt), and 9.2 mg (45%) of the by-product DOM-3,15-disulfate (as diammonium salt). By using LC-HR-MS, the following molecular masses were obtained. DOM-3-sulfate: 360.0880 g/mol (exact molecular mass: 360.0879 g/mol); DOM-15-sulfate: 360.0879 g/mol (exact molecular mass: 360.0879 g/mol); DOM-3,15-disulfate: 440.0448 g/mol (exact molecular mass: 440.0447 g/mol). Results of NMR measurements are summarized in the electronic supplementary material; NMR and

LC-HR-MS/MS spectra are given in Figures S2 and S3. The structures of DON- and DOM-sulfates are provided in Figure 1.

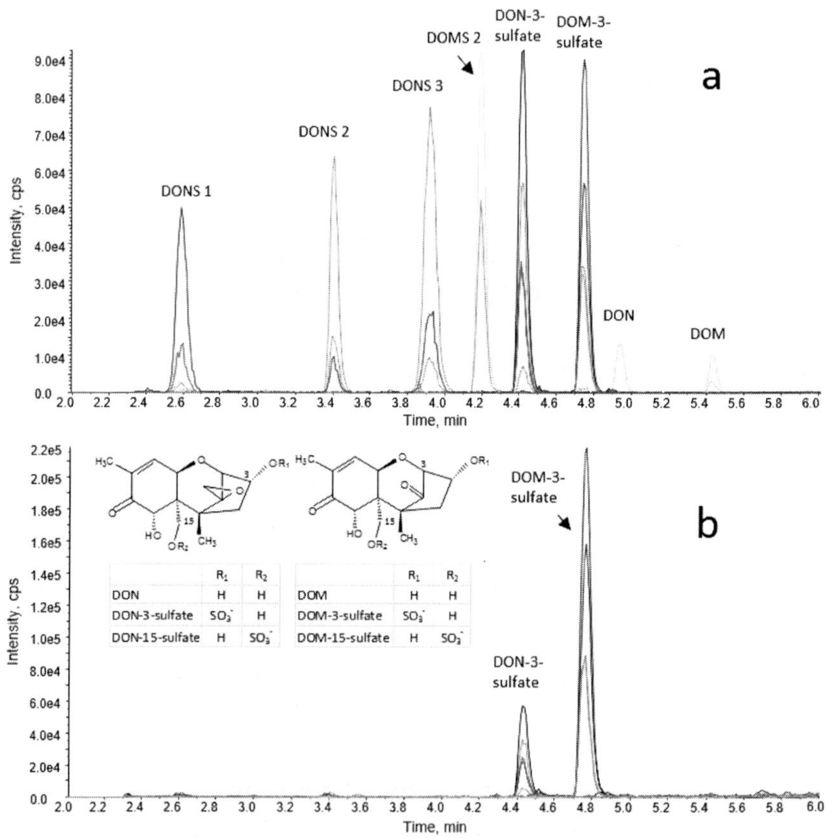

Figure 1. LC-MS/MS chromatogram of (**a**) a standard solution (30 ng/mL of DONS 1, DONS 2, DONS 3, DOMS 2, DON-3-sulfate, DOM-3-sulfate, DON, and DOM) and (**b**) of an excreta extract of a turkey fed a DOM-contaminated diet (1.6 mg/kg DOM and 0.2 mg/kg DON in feed).

2.2. Method Validation

For LC-MS/MS analysis, two LC gradient methods were developed, a long method capable of separating the 3/15 isomers of DON- and DOM-sulfate and a short method where the 3/15 isomers of DON- and DOM-sulfate are only partially separated. Chromatograms of a standard solution and of a turkey extract measured by the short method are shown in Figure 1, and a standard chromatogram recorded by the long method is depicted in Figure S4.

Prior to method validation, between four and six randomly selected samples of each matrix (excreta of turkeys, chickens, pullets, roosters, and chyme of pullets from both the DON- and DOM group, if available) were extracted and measured by the long gradient method. Due to presence of DON-3-sulfate and in part DOM-3-sulfate, but absence of the 15-sulfate isomer in all investigated samples, the long gradient method was validated only with respect to matrix effects, whereas the short gradient method was selected for full validation and later on for measurement of all samples.

Apparent recoveries (R_As), recoveries of extraction (R_Es), and matrix effects (SSE) for the short routine method are summarized in Table 1. Signal suppression and enhancement (SSE) of the long gradient method are given in Table S1 in the Supplementary Material. Recoveries of extraction ranged between 87% and 96% for all analytes. Hence, apparent recoveries were mainly determined by mass spectrometric SSE. Matrix effects for DON and DOM were between 80% and 108% in all sample extracts, resulting in apparent recoveries between 71% and 107%. Matrix effects for DON-3-sulfate, DOM-3-sulfate, as well as for DON- and DOM sulfonates were mostly between 100% and 120% with the exception of all analytes in chyme sample extracts and DON- and DOM sulfonates in excreta extracts of chickens. Accordingly,

R_{AS} mostly ranged between 90% and 110%. However, matrix enhancement by approximately 150% was observed for DON-3-sulfate, DOM-3-sulfate, and DONS 1 in jejunum and ileum of pullets. Exceptionally large matrix enhancement occurred for DONS 2, DONS 3, and DOMS 2 in chyme sample extracts and additionally for DONS 3 in excreta extracts of chickens. One reason is the lower dilution factor for chyme samples compared to excreta extracts, which was required because of lower analyte concentrations in jejunum and ileum compared to excreta. Additionally, DON sulfonates and DON-3-sulfate exhibit severe matrix enhancement in cereal sample extracts [32,33], and chyme samples still contain greater proportions of undigested cereal matrix. As indicated by these data, quantitation of DON- and DOM- sulfates and sulfonates in chyme and excreta samples of poultry requires determination of matrix effects in appropriately diluted blank extracts and correction by the respective R_{AS}.

Limits of detection (LODs) and quantitation (LOQs) are given in Table 2. In pure standard solution, LODs ranged between 0.3 (sulfates and sulfonates of the series 2 and 3) and 1.5 ng/mL (DON, DOM), LOQs between 1.0 and 4.5 ng/mL. LODs and LOQs in lyophilized samples depended on the dilution factor of the extract. In chyme samples, LODs were between 40 (sulfates and sulfonates of the series 2) and 200 (DOM) ng/g, LOQs between 120 and 600 ng/g. Highest LODs (up to 1000 ng/g for DOM) and LOQs (up to 3000 ng/g) were obtained for excreta of roosters, the extract of which was diluted 1 + 9 (v + v) before measurement.

Table 1. Validation of the short gradient method. R_A: apparent recovery, R_E: recovery of extraction, SSE: matrix effects.

		Dilution of Extract (v + v)	Average ± Standard Deviation (n = 3, Values in %)							
			DON-3-sulfate	DOM-3-sulfate	DON	DOM	DONS 1	DONS 2	DONS 3	DOMS 2
R_E	Excreta of turkeys		90 ± 1	91 ± 1	91 ± 1	89 ± 1	90 ± 2	90 ± 3	91 ± 3	90 ± 2
	Excreta of chickens		86 ± 5	88 ± 3	93 ± 4	87 ± 3	91 ± 3	96 ± 3	87 ± 7	90 ± 3
SSE	Excreta of turkeys	1 + 2	102 ± 1	99 ± 0	96 ± 2	80 ± 2	103 ± 2	122 ± 3	105 ± 1	107 ± 1
	Excreta of chickens	1 + 2	117 ± 2	112 ± 3	117 ± 2	97 ± 1	121 ± 1	138 ± 1	208 ± 6	131 ± 1
	Jejunum of pullets	1 + 1	153 ± 1	150 ± 1	93 ± 1	107 ± 1	149 ± 2	229 ± 14	334 ± 25	242 ± 20
	Ileum of pullets	1 + 1	149 ± 7	138 ± 1	99 ± 1	107 ± 1	177 ± 2	292 ± 9	365 ± 9	256 ± 6
	Excreta of pullets	1 + 5	100 ± 1	110 ± 1	101 ± 0	101 ± 0	121 ± 3	115 ± 1	110 ± 2	115 ± 0
	Excreta of roosters	1 + 9	93 ± 0	110 ± 2	101 ± 1	108 ± 1	133 ± 2	116 ± 2	116 ± 0	107 ± 1
R_A	Excreta of turkeys	1 + 2	92 ± 3	90 ± 2	87 ± 3	71 ± 3	92 ± 2	111 ± 1	96 ± 1	96 ± 2
	Excreta of chickens	1 + 2	98 ± 6	98 ± 6	107 ± 6	84 ± 4	110 ± 4	132 ± 7	178 ± 15	118 ± 2
	Jejunum of pullets	1 + 1	135 ± 2	134 ± 3	86 ± 2	94 ± 1	135 ± 1	213 ± 1	297 ± 5	218 ± 1
	Ileum of pullets	1 + 1	131 ± 1	124 ± 1	91 ± 1	94 ± 1	160 ± 2	272 ± 13	325 ± 22	230 ± 18
	Excreta of pullets	1 + 5	88 ± 6	98 ± 1	93 ± 1	89 ± 1	110 ± 2	107 ± 8	98 ± 8	104 ± 5
	Excreta of roosters	1 + 9	82 ± 1	98 ± 1	93 ± 0	95 ± 0	120 ± 3	108 ± 1	103 ± 2	96 ± 0

Table 2. Limits of detection (LODs) in pure standard solution and in freeze-dried chyme and excreta samples of poultry. Limits of quantitation (LOQs) were by the factor of 3.3 higher.

Matrix	DON-3-sulfate	DOM-3-sulfate	DON	DOM	DONS 1	DONS 2	DONS 3	DOMS 2
Pure standard solution (ng/mL)	0.3	0.3	1.5	1.5	0.6	0.3	0.3	0.3
Excreta of turkeys (ng/g)	59	98	196	300	119	59	68	59
Excreta of chickens (ng/g)	51	98	196	400	119	59	68	69
Jejunum of pullets (ng/g)	40	40	130	200	79	40	46	40
Ileum of pullets (ng/g)	40	40	130	200	79	40	46	40
Excreta of pullets (ng/g)	102	147	391	800	237	119	137	138
Excreta of roosters (ng/g)	198	245	652	1000	396	198	228	198

Calibration functions in neat standard solution were linear from the LOQ up to at least 674 ng/mL for all analytes. Matrix-matched calibration functions were linear in the investigated calibration ranges (up to 674 ng/mL for excreta extracts of turkeys and chickens; up to 600 ng/mL for chyme and excreta extracts of pullets and excreta extracts of roosters) with the following exceptions: calibration curves of DONS 2, DONS 3, and DOMS 2 were linear up to 300 ng/mL in chyme extracts, and calibration functions of DONS 3 were linear up to 337 ng/mL in excreta extracts of chickens.

The relative standard deviation (RSD) of triplicate work-up and measurement of blank turkey and chicken excreta samples spiked at 4 concentration levels on one day was below 10% for each substance at each concentration level (mostly between 2% and 5%). The inter-day RSD of work-up and analysis of one freeze-dried excreta sample of one turkey of the DOM group containing 834 ng/g DON-3-sulfate and 3420 ng/g DOM-3-sulfate was 11% and 9% for DON-3-sulfate and DOM-3-sulfate, respectively (n = 10). The inter-day RSD of sample preparation and measurement of one broiler excreta sample containing 7720 ng/g DON-3-sulfate that was worked-up and analysed on 10 different days was 5%.

2.3. Analysis of Samples

Excreta and chyme samples of four different feeding trials with different poultry species were carried out in the past in order to study the metabolization of DON and DOM, to assess biological recoveries, and to investigate the effects of DON on health and performance in poultry. These samples were re-analysed using the newly developed LC-MS/MS method, which included DON- and DOM-3-sulfate as well as DON- and DOM sulfonate metabolites. An overview of these four feeding trials is given in Table 5.

Turkey and chicken trial: The two feeding trials with turkeys and chickens were carried out in order to study the metabolization of DON and DOM in poultry and in order to assess the biological recoveries of these two mycotoxins after oral administration at concentrations relevant in practice. For that, the mycotoxin content in poultry feed (between 1.5 and 1.7 mg/kg) was kept well below the guidance value of 5 mg/kg in complementary and complete feeding stuffs. As birds excrete a mixture of white pasty urine and feces via the cloaca, only the combined excreta samples were collected.

Both in turkeys and in chickens, DON-3-sulfate was the major DON metabolite, followed by DON. Analogous to that, DOM-3-sulfate turned out to be the major metabolite of orally administered DOM. DOM itself could not be detected in any of the excreta samples of the turkey and the chicken trial, nor could any of the DON- or DOM sulfonates. The average DON equivalent concentrations of the main DON- and DOM metabolites in excreta of turkeys and chickens (n = 4 for each species) at the individual sampling times are given in Figure 2. In lyophilized excreta of the individual turkeys of the DON group, maximum DON equivalent concentrations of DON-3-sulfate ranged from 8.8 to 16 μg/g and were obtained between 2 h (mean concentration 9.6 μg/g) and 4 h (mean concentration 8.8 μg/g) after provision of feed. Maximum DON equivalent concentrations of DOM-3-sulfate in excreta of turkeys of the DOM group were measured in excreta collected 2 h after start of feeding and ranged between 5.3 and 18 μg/g (average 13 μg/g). Due to the presence of 0.3 mg/kg DON in feed of the DOM group, DON-3-sulfate was also excreted by turkeys of the DOM group. The excretion pattern was similar to that of DOM-3-sulfate, with a maximum mean value of 2.5 μg/g in DON equivalents. A chromatogram of a turkey excreta extract of the DOM group is given in Figure 1.

In freeze-dried excreta of chickens, average DON equivalent concentrations of DON-3-sulfate in the DON group and of DOM-3-sulfate in the DOM group were 2.8 and 3.0 µg/g, respectively, 3 h after provision of feed. At the later sampling points (6, 9, 12 and 24 h after start of feeding), average DON equivalent concentrations of DON-3-sulfate increased to between 4.8 and 5.8 µg/g in the DON group and mean DON equivalent concentrations of DOM-3-sulfate were between 4.0 and 4.8 µg/g in the DOM group. The average DON equivalent concentration of DON-3-sulfate in excreta of the control group was between 0.8 and 1.3 µg/g.

The feed consumption by turkeys was not significantly different for the three feeding groups (DON, DOM, control) and ranged from 110 to 380 g per turkey on the day of the feeding trial. Likewise, feed consumption in chickens was similar across the three feeding groups with average values between 105 and 115 g per chicken and day. In Table 3, the average amounts of DON-3-sulfate, DOM-3-sulfate and DON excreted by turkeys and chickens between 2 h and 24 h after the start of feeding are given. Comparison of DON equivalent amounts of total excreted metabolites and ingested toxins yielded biological recoveries in turkeys between 34% and 71% (52% ± 16%, average ± standard deviation, $n = 4$) in the DON group, between 33% and 76% (51% ± 18%) in the DOM group, and between 68% and 75% (72% ± 3%) in the control group. In chickens, biological recoveries ranged between 69% and 95% (80% ± 13%) in the DON group, between 68 and 78% (74% ± 4%) in the DOM group, and between 93% and 126% (106% ± 14%) in the control group.

Pullet trial: The original aim of the feeding trial with pullets was to investigate interactions between *Ascaridia galli* infection of pullets and DON contamination of feed [28]. As DON-3-sulfate and DOM-3-sulfate had not been suspected as major DON metabolites in poultry at that time, biological recoveries had been poor (≤5%). Therefore, the excreta samples obtained in this animal experiment were re-analysed by the newly developed LC-MS/MS method including DON- and DOM-sulfate and sulfonate metabolites. As contents of ileum and jejunum were also collected, the metabolite pattern across the pullets' GI tract could be studied as well.

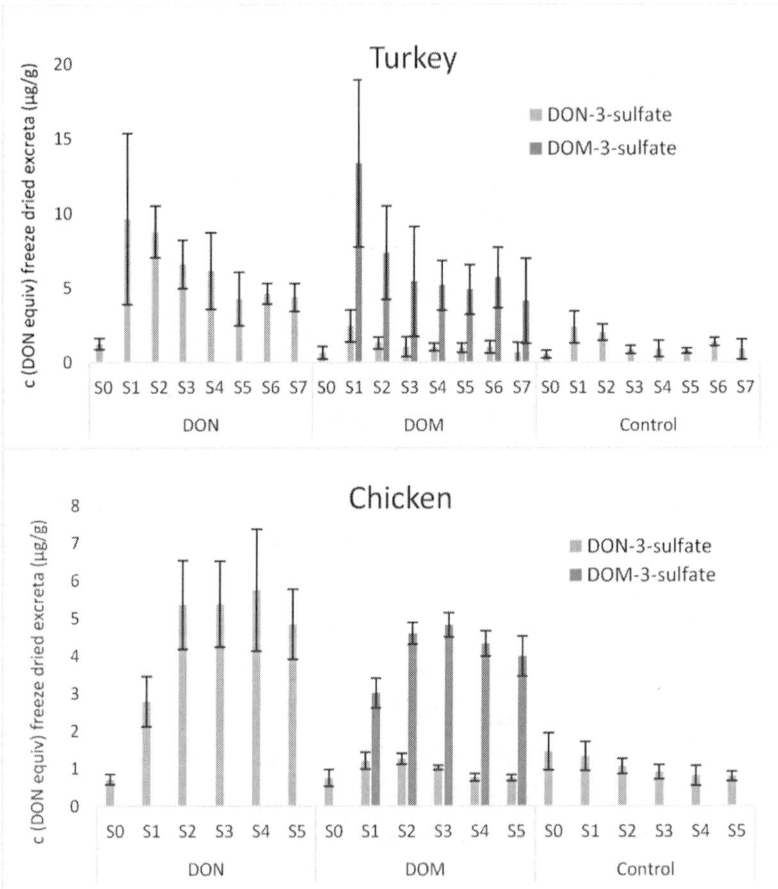

Figure 2. Average DON equivalent concentrations ± standard deviation (n = 4) of DON-3-sulfate and DOM-3-sulfate in excreta of turkeys and chickens at different sampling times (turkeys: 0, 2, 4, 6, 8, 11, 15 and 24 h after start of feeding; chickens: 0, 3, 6, 9, 12 and 24 h after start of feeding).

Table 3. Amounts of DON- and DOM metabolites excreted by turkeys and chickens between 2 h and 24 h after start of feeding (average ± natural standard deviation, n = 4, values in μg DON equivalents).

	Group	DON-3-sulfate	DOM-3-sulfate	DON	Sum Excreted	Consumed *	Biological Recovery (%)
Turkey	DON	144 ± 40	0.7 ± 0.8	5.2 ± 1.2	150 ± 42	304 ± 96	52 ± 16
	DOM	27 ± 13	149 ± 67	1.0 ± 1.4	177 ± 80	359 ± 159	51 ± 18
	Control	29 ± 7	0 ± 0	0 ± 0	29 ± 7	40 ± 10	72 ± 3
Chicken	DON	450 ± 112	0 ± 0	17 ± 12	476 ± 118	591 ± 77	80 ± 13
	DOM	72 ± 5	321 ± 30	1 ± 3	397 ± 33	536 ± 17	74 ± 4
	Control	71 ± 15	0 ± 0	0 ± 0	71 ± 15	67 ± 5	106 ± 14

* The average amounts of DON and DOM consumed were calculated by multiplying the concentration in feed by the amount of feed consumed by the individual animals between start and end of feeding.

The average concentrations of DON metabolites in chyme and excreta samples are given in Table 4. Again, DON-3-sulfate was the main DON metabolite. However, contrary to chicken of the feeding trial described above, DOM-3-sulfate was also present in ileum and excreta samples of pullets. In addition, DON sulfonate metabolites of the series 1, 2 and 3 could be detected and quantified in jejunum and ileum samples. In general, both the metabolite concentrations and the metabolite pattern changed across the GI tract. In jejunum, DON-3-sulfate concentrations in the DON groups with and without *A. galli* infection were on average 3.5 times lower than in ileum and 60 times lower than in excreta samples. DOM-3-sulfate concentrations also increased from jejunum to excreta, reaching about 12%–15% of the DON-3-sulfate concentration in the latter matrix. DON sulfonate 2 (DONS 2) and DON sulfonate 3 (DONS 3) could be quantified in jejunum of the DON groups. In ileum, the concentration of DONS 2 remained similar, but DONS 3 could not be detected, most likely due to its instability under alkaline conditions [24,34]. However, traces of DONS 1 were observed in ileum, which might result from decomposition of DONS 3. In excreta, limits of quantitation were higher because of measurement of more diluted samples, so that only traces of DONS 2 were detected.

Biological recoveries reached values of 151% ± 5% in the DON group without *A. galli* infection and 131% ± 11% in the DON group infected by *A. galli* (see Table S2 in the Supplementary Material). Interestingly, statistical analysis revealed significant differences both between the concentrations and between the excreted amounts of DON-3-sulfate and DOM-3-sulfate in excreta of pullets of the DON groups with and without infection by *A. galli*. Excreta of pullets not infected by *A. galli* showed significantly greater concentrations of both sulfate metabolites than excreta of pullets infected with worms. Likewise, pullets not infected by *A. galli* excreted significantly greater amounts of DON-3-sulfate and of DOM-3-sulfate than pullets infected with *A. galli*.

Table 4. Average concentrations of DON metabolites in chyme and excreta samples of pullets receiving 65 g feed containing 4.4 mg/kg DON (corresponding to 286 μg DON) per day. CON: negative control, -: no *A. galli* infection, +: *A. galli* infection, n.d.: not detected, tr: traces.

Matrix	Group	*n*	c (Average ± std dev, μg/g Freeze Dried Sample in DON Equivalents)					
			DON-3-Sulfate	DOM-3-Sulfate	DON	DONS 1	DONS 2	DONS 3
Jejunum	CON -	3	n.d.	n.d.	n.d.	n.d.	n.d.	n.d.
	CON +	3	tr *	n.d.	n.d.	n.d.	n.d.	n.d.
	DON -	3	0.34 ± 0.04	n.d.	tr	n.d.	0.20 ± 0.05	0.08 ± 0.02
	DON +	3	0.35 ± 0.04	n.d.	tr	n.d.	0.17 ± 0.00	0.12 ± 0.04
Ileum	CON -	2	tr	n.d.	n.d.	n.d.	n.d.	n.d.
	CON +	3	tr	n.d.	n.d.	n.d.	n.d.	n.d.
	DON - **	2	1.26 ± 0.53	tr	tr	tr	0.10 ± 0.01	n.d.
	DON +	3	1.09 ± 0.11	1.00 ± 1.38	tr	tr	0.20 ± 0.12	n.d.
Excreta	CON -	9	1.83 ± 0.34 [a]	tr	n.d.	n.d.	n.d.	n.d.
	CON +	9	2.00 ± 1.00 [a]	tr	n.d.	n.d.	n.d.	n.d.
	DON -	9	22.9 ± 0.8 [b]	3.47 ± 0.14 [a]	tr	n.d.	tr	n.d.
	DON +	9	20.3 ± 1.8 [c]	2.49 ± 1.02 [b]	tr	n.d.	tr	n.d.

*: tr in one sample of three. **: for groups with $n = 2$, the value is given as average ± (max-min)/2. Statistical analysis (excreta): mean values with different superscripts within one column are significantly different.

Roosters: Excreta samples of roosters administered feed contaminated with 11 mg/kg DON [29] were analysed in order to study the metabolite pattern in roosters. Average DON-3-sulfate concentrations were 29 ± 3 µg/g freeze-dried excreta in DON equivalents. Surprisingly, mean DOM-3-sulfate concentrations amounted to 15 ± 3 µg/g in DON equivalents, yielding a DON-3-sulfate to DOM-3-sulfate ratio of 1.9 to 1. DONS 2 occurred at 0.8 ± 0.1 µg/g in DON equivalents, whereas only traces of DON were detected. Other compounds were below the respective LODs in rooster excrement extracts (see Table 2).

2.4. Toxicity assessment of DOM-3-Sulfate and DOM-15-Sulfate

The toxicity of DON, DON-3-sulfate, and DON-15-sulfate on the ribosome had recently been assessed in *in vitro* transcription/translation assays with wheat germ extract [31]. In short, DON-3-sulfate did not show any toxicity in the investigated concentration range, whereas DON-15-sulfate was *ca.* 44 times less toxic than DON. To compare the toxicities of DON, DOM, DOM-3-sulfate, and DOM-15-sulfate, *in vitro* transcription/translation assays were performed with mammalian ribosomes (Figure 3). While 50% reduction of *in vitro* translation by rabbit reticulocyte lysate was obtained at 0.6 µM DON, the IC_{50} value of DOM was 410 µM. DOM-3-sulfate and DOM-15-sulfate did not significantly inhibit *in vitro* translation at concentrations up to 100 µM. The translation efficiency was still 81% and 75% at 500 µM DOM-3-sulfate and DOM-15-sulfate, respectively. Inhibition of translation was not significantly different for DOM-3-sulfate and DOM-15-sulfate at substrate concentrations up to 250 µM. At a concentration of 500 µM, differences in translation were minor, but significant. Based on IC_{20} values for DON (0.25 µM), DOM (122 µM), DOM-3-sulfate (490 µM), and DOM-15-sulfate (420 µM),

DOM-3-sulfate is less toxic than DOM by a factor of four and less toxic than DON by almost a factor of 2000. Similarly, DOM-15-sulfate shows lower toxicity on the ribosome than DOM by a factor of 3.4 and lower toxicity than DON by a factor of nearly 1700. Statistical analysis revealed significantly greater inhibition of translation by DOM compared to DOM-3-sulfate and DOM-15-sulfate from 250 μM onwards. These results confirm that de-epoxidation of DON leads to a less toxic metabolite and that sulfation of trichothecenes is, similar to glucosylation in plants, a detoxification mechanism.

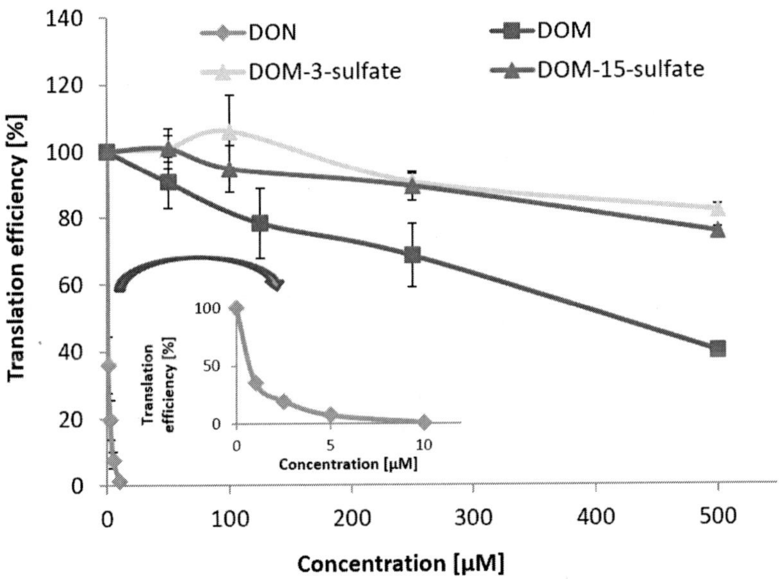

Figure 3. Comparative toxicity of DON, DOM, DOM-3-sulfate, and DOM-15-sulfate on the ribosome, determined by an *in vitro* transcription/translation assay with rabbit reticulocyte lysate. Error bars denote the analytical standard deviation of three independent determinations.

3. DISCUSSION

Concluding from our results and those published in the most recent literature [25,26], DON-3-sulfate and DOM-3-sulfate are the missing link to quantitative biological recoveries of DON in poultry. Similar to previous studies reporting percentages of DON in excreta as being between 2% and 6% of total ingested DON [28,35], excreted DON was 2% and 3% of ingested DON in our turkey and chicken trial and approximately 4% in the pullet experiment. DOM was rarely detected in excreta samples in the literature. Whereas Wan *et al.* and Awad and co-workers could not detect DOM in excreta extracts [25,35], Dänicke *et al.*, applying immunoaffinity clean-up as sample preparation, determined 0.5 µg DON/kg b.w./day for pullets consuming *ca.* 300 µg DON/kg b.w./day, corresponding to approximately 0.2% of ingested DON [28]. Re-analyzing these pullet excreta samples, we could not detect DOM, most likely due to higher LODs due to greater dilution factors (see below). Similarly, DOM could not be detected in any of the excreta samples of our turkey and chicken experiments. One reason might be lack of microbes capable of de-epoxidizing DON. Another reason is that our methods were not designed to detect trace concentrations of DOM, but to allow co-determination of DON- and DOM-sulfates, DONS, DOMS, DON and DOM. Separation of DONS and DOMS requires formic acid as a mobile phase additive [24]. However, use of formic acid resulted in fivefold higher LODs for DON and DOM compared to acetic acid. Still, this compromise had to be made for proper separation of DON sulfonates. LOQ values for DON and DOM in a pure standard solution (5 ng/mL) were higher than those obtained by Devreese *et al.*(0.1 and 0.5 ng/mL for DON and DOM), but similar to those obtained in our previous work on DON- and DOM sulfonates (2 ng/mL on 5500 QTrap, 5 ng/mL on 4000 QTrap). LOQ values in lyophilized pullet feces obtained by Dänicke and co-workers were 0.8 and 1.6 ng/g [28], and as such far better than ours. The reason for this great difference is that Dänicke *et al.* used multi-step sample preparation including defatting and immunoaffinity clean-up. In our work, on the contrary, great dilution factors were chosen on purpose due to the high concentrations of DON-3-sulfate and in order to minimize matrix effects.

As DOM and DOM-sulfates are much less toxic than DON in an *in vitro* translation assay using animal ribosomes, and the main mode of action of all trichothecenes is the suppression of translation, these metabolites can be considered as detoxification products of DON. In the following, the question about location of formation, absorption and elimination of DON-3-sulfate and DOM-3-sulfate will be tackled by putting our findings and recent literature results in context with older studies on DON metabolization in poultry. For this purpose, it is important to note that poultry excrete white pasty urine into the cloaca where it is mixed with solid excrements from the GI tract. Hence, administration of native DON to poultry and collection of excrements does not give information about the origin of excreted DON-3-sulfate and DOM-3-sulfate. Another peculiarity of poultry is that, in addition to the hepatic portal system, a renal portal system exists. Both systems drain the intestine via the *Vena mesenterica cranialis* and *V. mesenterica caudalis*, respectively, and might contribute to renal and/or hepatic metabolism and first pass elimination as suggested by Rotter *et al.* [36].

As early as in 1986, Prelusky and co-workers administered a single oral dose of ^{14}C-labeled DON to white Leghorn hens by crop intubation [37]. They observed low levels of radioactivity in plasma and reported the total body burden (excluding bile, GI tract and excreta) of radioactively-labeled DON and its metabolites to be less than 2%. The highest accumulation of radioactivity was determined in bile 6 h after treatment. Within 24 h, 79% of radioactivity was excreted. These results suggested low systemic absorption of DON and/or its metabolites and significant biliary excretion.

Two years later, Lun *et al.* [38], administering native DON to laying domestic chickens, reported that DON disappeared from the GI tract between the crop and the jejunum. However, only low levels of DON were detected in portal vein blood. Considering little decrease of DON when incubated with gastric fluid or juice from the small intestine [38], they set up the hypothesis of rapid post-absorptive modification of DON by enterocytes and hepatic elimination into bile. In the same study, *in vitro* incubation of native DON with contents of colon and caecae resulted in the reduction of the DON concentration by factors of 3, 9 and 20 after 6, 12 and 24 h of incubation, respectively.

One year later, Lun and co-workers administered tritium labeled DON by crop intubation to colostomized white Leghorn hens [39]. Interestingly, 68% of the administered radioactivity was excreted within 24 h into urine, whereas only 6% was eliminated via feces. Radioactivity in systemic blood accounted for only 7% of the administered radioactivity. Radioactivity along the GI tract of intact crop intubated hens decreased rapidly after the crop, but increased again in the large intestine six hours after administration, which is most likely due to retrograde movement of urine from cloaca into colon. However, contrary to observations by Prelusky and co-workers [37], biliary excretion of radioactivity was <1%. These results confirmed the hypothesis set up by the same working group in 1988 that ingested DON is rapidly absorbed between crop and upper jejunum and metabolized shortly after absorption [38]. Yet, extensive excretion of the altered DON via bile was not confirmed as most of the radioactivity was excreted into urine.

Results obtained in more recent studies where native or radiolabeled DON was administered to poultry mostly confirmed and complemented the earlier published data. Yunus *et al.* [27]

recovered only 0.04% of total DON fed to broilers 1 h after oral treatment in plasma in the form of DON and DOM. Likewise, Dänicke and co-workers determined only trace levels of DON in plasma of pullets fed with poultry feed containing 4.4 mg/kg DON. However, two recent studies reported an oral bioavailability of 20% for DON administered to broiler chickens [40], and 21% for DON fed to turkeys [26]. Higher values than previously reported might be due to different ways of calculating the oral bioavailability. In accordance with earlier studies, rapid absorption, fast elimination, and extensive biotransformation were reported [26,40]. Comparison of oral and IV application of DON and inclusion of DON-3-sulfate in the analytical method for semiquantitative determination yielded precious information on the toxicokinetic behavior of DON and on the location of sulfation in turkeys and broiler chickens [26]. Already 5 min after oral administration, DON and DON-3-sulfate were detected in plasma. Maximum values of DON were reached 10 min post dosing, and maximum values of DON-3-sulfate between 20 and 30 min after application. Only 5 min after IV administration of DON, the average peak area ratio of DON-3-sulfate to DON was 10 in turkeys and 243 in broiler chickens. This finding points to very rapid sulfation, with the liver, extrahepatic tissues, and the intestinal mucosa being potential metabolism sites. In addition, average peak area ratios of DON-3-sulfate to DON were much greater after oral, rather than after IV administration in both avian species. This observation strongly substantiates the hypothesis by Lun *et al.* [38,39] that sulfation is already performed in enterocytes of the intestinal mucosa.

Our current turkey and chicken trial revealed that metabolization of DON and DOM was very similar in both avian species. Excretion kinetics could not be established because animals had *ad libitum* access to DON and DOM contaminated feed during the day. Despite large inter-individual differences between the turkeys of one feeding group, maximum metabolite concentrations in excreta between 2 and 4 h after the start of feeding and maximum feed intake in the morning suggest fast transit times in turkeys. In chickens, excreta were sampled 3 and 6 h after provision of DON or DOM contaminated feed, so that transit times can only roughly be estimated. Still, maximum concentrations obtained at the latest 6 h after start of feeding corroborate fast excretion of DON- and DOM metabolites also in chickens.

Low concentrations of DON-3-sulfate and DON in chyme samples (content of jejunum and ileum) of pullets support the hypothesis of rapid and extensive absorption of DON from the small intestine [39]. The origin of DON-3-sulfate in jejunum and ileum might be biliary excretion of DON-3-sulfate formed in enterocytes or in the liver. As chyme samples were collected 3 h after administration of DON, biliary excretion might not have arrived at its maximum which was reported to occur *ca.* 6 h after dosing [37].

Presence of DOM-3-sulfate in excreta of pullets and roosters might be due to formation of DOM by intestinal microbes, absorption of DOM from the GI tract, sulfation in the intestinal mucosa, liver and/or kidney, and excretion into the cloaca via urine or back into the GI tract via bile. The latter could also explain low concentrations of DOM-3-sulfate in ileum. An additional hypothesis is antiperistaltic retrograde movement of DON-3-sulfate from the cloaca to the microbial rich caecae and microbe-mediated formation of DOM-3-sulfate from DON-3-sulfate. A third possibility might be sulfation in DOM by intestinal microbes. Absence of DOM-3-sulfate in excreta samples of our turkey and chicken trial might be due to differences in the intestinal microflora, specifically due to lack of microbes capable of de-epoxidizing DON. Although infection with *A. galli* resulted in significantly lower formation of DON-3-sulfate and DOM-3-sulfate in pullets, the formation pattern of DON metabolites was unaltered by worm infection.

Concluding from the DON sulfonate pattern across the GI tract of pullets, DON sulfonates are formed early in the digestion process. The mechanism of DON sulfonate formation is currently unknown. Some experiments aiming at elucidating the mechanism of DON-, DOM-, and DON-3-glucoside (D3G) sulfonate formation in rats had been carried out lately [24], but were inconclusive. However, because the major part of the sulfonates was recovered in feces of DON treated rats and the formation pattern shifted to DOM sulfonate on the second day after treatment, authors speculated that DON-, DOM-, and D3G sulfonates might be formed in a Michael addition with reactive inorganic sulfur species possibly present in intestinal chyme rather than by liver enzymes.

Biological recoveries of DON in the current study are much higher than those previously published for native DON administered to chickens, which were ≤5% [28]. However, they compare nicely to values published for experiments with radiolabeled DON. Wan *et al.* [25], using radioisotope counting radio-HPLC, reported biological recoveries of 81%, 89% and 95% in the time span of 24, 48 and 72 h after oral administration of a single dose of radiolabeled DON. Likewise, Prelusky *et al.* [37] recovered 79%, 92% and 99% of radiolabeled DON by 24, 48 and 72 h after single bolus administration to white Leghorn hens. The finding that excretion is not complete 24 h after treatment explains incomplete biological recoveries in our feeding trials with turkeys and chickens. Lower biological recoveries in turkeys than in chickens hint at delayed excretion in turkeys compared to chickens. Recoveries greater than 100% in the pullet trial might be due to presence of DON-3-glucoside (D3G) in the Fusarium contaminated diet used in the animal experiment. As D3G was not determined in the administered feed, its presence and subsequent *in vivo* metabolization to DON- and DOM-3-sulfate cannot be excluded.

To summarize, our findings and data from the recent literature substantiate, complement, and extend results and hypotheses from earlier published articles on the absorption, metabolism, and excretion of DON in poultry. DON is, to a great extent, rapidly absorbed from the GI tract between crop and jejunum. After absorption, DON is extensively metabolized to DON-3-sulfate in the intestinal mucosa, the liver, and possibly in the kidney. DON-3-sulfate is rapidly and completely eliminated into bile by the liver and/or into urine by the kidney. DOM-3-sulfate is an important natural DON metabolite in pullets and roosters and might originate from sulfation of microbe-formed DOM by intestinal mucosa, liver, and/or kidney or from microbe mediated deepoxidation of urinary DON-3-sulfate transported into caecae by retrograde movement.

4. EXPERIMENTAL SECTION

4.1. Reagents

Methanol (MeOH, LC gradient grade) and formic acid (98%–100%, p.a.) were purchased from Merck (Darmstadt, Germany). Formic acid (LC-MS gradient grade) for LC-MS/MS was obtained from Sigma Aldrich (Vienna, Austria). Reagents for synthesis of DOM-sulfates were purchased from ABCR (Karlsruhe, Germany) and Sigma Aldrich (Vienna, Austria). In all experiments, ultra-pure water (delivered by a Purelab Ultra system (ELGA LabWater, Celle, Germany)) was used.

4.2. Synthesis, Purification and Characterization of DOM-3-Sulfate and DOM-15-Sulfate

DOM as the starting material for synthesis of DOM-3-sulfate and DOM-15-sulfate was produced from DON as described earlier [24] and purified by preparative chromatography. The preparative HPLC system and column were the same as described in [34]. Mobile phase A consisted of methanol/water/formic acid (10/89.9/0.1, v/v/v), mobile phase B of methanol/formic acid (99.9/0.1, v/v). Gradient elution started at 0% B from 0 to 1 min and continued with linear increase to 100% B from 1 to 6 min. After 1 min at 100% B the starting conditions were re-established from 7 to 7.1 min and the column was re-equilibrated at 0% B until 10 min. The injection volume was 700 μL, the flow rate 16 mL/min and DOM was detected by measurement of UV absorption at 220 nm. The collection window of DOM was 5.66–6.03 min.

The synthesis of DOM-3-sulfate and DOM-15-sulfate was carried out as described for DON [31] with minor modifications. In short, it included synthesis of protected intermediates (4.2.1) and subsequent deprotection (4.2.2) followed by purification by column chromatography. The progress of all reactions was monitored by thin layer chromatography (TLC) using silica gel 60 F_{254} TLC plates (Merck, Germany). All chromatograms were visualized using ceric ammonium molybdate/Hanessian's stain in EtOH/sulphuric acid. Column chromatography was performed on silica gel 60 (40–63 μm) using a Sepacore™ Flash System (Büchi, Switzerland) or small glass columns.

4.2.1. Synthesis of the Protected Intermediates

DOM (140.0 mg, 0.50 mmol, 1.0 equivalents (eq.)) was dissolved in 20 mL of dry dichloromethane (DCM), cooled to 0 °C, and 1,2-

dimethylimidazole (144.0 mg, 1.50 mmol, 3.0 eq.) was added to the mixture. Finally, 2,3-dimethyl-1-((2,2,2-trichloroethoxy) sulfonyl)-1H-imidazolium trifluoromethanesulfonate (342.8 mg, 0.75 mmol, 1.5 eq.) was added and the reaction was allowed to reach room temperature overnight. TLC after 1 day showed nearly full conversion of the starting material. Hence, the reaction was directly subjected to column chromatography (hexane/ethyl acetate 3/1, v/v) upon which three protected intermediates (2,2,2-trichloroethyl-DOM-3-sulfate (71.0 mg, 29%), 2,2,2-trichloroethyl-DOM-15-sulfate (14 mg, 6%) and bis(2,2,2-trichloroethyl) DOM-3,15-disulfate (30.0 mg, 9%)) were obtained.

4.2.2. Deprotection of Protected Intermediates

Deprotection of the protected intermediates was carried out as described in [31]. The starting material was dissolved in MeOH (1 mL/10 μmol). Ammonium formate (3 eq. for the monosulfates, 6 eq. for the disulfate) as well as Zn dust (9 and 18 eq., respectively) were added, and the reaction mixture was placed in an ultrasonic bath at room temperature. The progress of the reaction was monitored by TLC until substantial amounts of products were formed (20 to 120 min). After filtration through celite, the remaining residue was subjected to column chromatography to end up with the corresponding sulfates as ammonium salts. For this purpose, a mobile phase of DCM/MeOH/NH_4OH (10/4/1, v/v/v) was used. The final products were dissolved in water, lyophilized, and finally obtained as a white powder.

4.2.3. NMR Spectroscopy

^1H and ^{13}C spectra were recorded on a Bruker Avance DRX-400 MHz and a Bruker Avance III 600 MHz spectrometer (Bruker,

Germany). Data were recorded and evaluated using TOPSPIN 1.3 and TOPSPIN 3.2 (Bruker Topspin). All chemical shifts are given in ppm relative to tetramethylsilane. The calibration was done using residual solvent signals. Multiplicities are abbreviated as s (singlet), d (doublet), t (triplet), q (quartet) and b (broad signal). Deuterated solvents were purchased from Eurisotop (Gif sur Yvette Cedex, Paris, France).

4.2.4. LC-HR-MS/MS

LC-HR-MS(/MS) spectra were recorded on a 6550 iFunnel Q-TOF instrument coupled to a 1290 Infinity UHPLC system (both Agilent Technologies, Waldbronn, Germany). Chromatographic separation was carried out on a Zorbax Eclipse Plus C18 Rapid Resolution High Definition column (2.1 × 150 mm, 1.8 μm particle size, Agilent, Waldbronn, Germany) at a flow rate of 0.25 mL/min using gradient elution (0 min: 15% B, 6 min: 100% B, 7 min: 100% B, 7.1 min: 15% B, 9 min: 15% B). Mobile phase A was water/formic acid (99.9/0.1, v/v), mobile phase B MeOH/formic acid (99.9/0.1, v/v). Compounds were ionized by electrospray ionization in the negative mode and measured first in full scan and then in targeted MS/MS mode at a collision energy of 30 eV (both in the range from*m/z* 40-1000). Electrospray ionization was carried out at a gas temperature of 140 °C, drying gas flow of 14 L/min, nebulizer pressure of 35 psig, sheath gas temperature of 350 °C, and sheath gas flow of 11 L/min. The capillary voltage was 4500 V, the nozzle voltage 300 V. Data acquisition was achieved in the 2 GHz extended dynamic range mode.

4.3. Standards and Standard Solutions

Solid DON (purity >95%) as well as standard solutions of DON and DOM (both 50 mg/L in acetonitrile) were supplied by Romer Labs GmbH (Tulln, Austria). DON-3-sulfate and DON-15-sulfate (both 95% purity) were synthesized according to [31]. DOM, DOM-3-sulfate, and DOM-15-sulfate were produced as described above (4.2). DON sulfonates 1, 2, 3 and DOM sulfonate 2 were produced as described in [24].

A mixed standard solution containing 300 mg/L of DON-3-sulfate, DOM-3-sulfate, DON, DOM, DONS 1, DONS 2, DONS 3, and DOMS 2 was prepared in methanol/water/formic acid (20/79.9/0.1, v/v/v). This solution and several dilutions thereof were used for spiking experiments and the establishment of pure solvent and matrix-matched calibration functions.

4.4. Sample Preparation Methods for Determination of DON, DOM, and Their Sulfates in Excreta and GI Samples of Poultry

Extraction of DON, DOM and their sulfate- and sulfonate metabolites from excreta samples was carried out according to a procedure previously optimized for the extraction of DON- and DOM sulfonates from rat feces [24]. A 300 mg aliquot of homogenized lyophilized excreta or GI content sample was consecutively extracted with 4, 3 and 3 mL of methanol/water/formic acid (49.5/49.5/1, v/v/v) by shaking for 30, 20 and 10 min in 15 mL polypropylene tubes. Prior to LC-MS/MS analysis, aliquots of the pooled extracts were diluted with water (turkeys, chickens: 1 + 2; pullets: jejunum and ileum: 1 + 1, excreta: 1 + 5; roosters: 1 + 9) and centrifuged at 14,000 × *g*.

4.5. LC-MS/MS Analysis of DON, DOM, Their Sulfates and Their Sulfonates in Excreta of Poultry

LC-MS/MS analyses were carried out on an Agilent 1290 series UHPLC system coupled to a 6500 QTrap mass spectrometer equipped with an IonDrive Turbo V® source (Sciex, Foster City, CA, USA). Chromatographic separation was achieved on a Kinetex Biphenyl column (150 × 3 mm, 2.6 µm) protected by a SecurityGuard ULTRA pre-column of the same stationary phase (both Phenomenex, Aschaffenburg, Germany) at 30 °C and at a flow rate of 0.4 mL/min. Mobile phase A consisted of water/formic acid, mobile phase B of methanol/formic acid (both 99.9/0.1, v/v). Two different gradient methods were used, a short one for routine measurements, and a long one for separation of the 3/15 isomers of DON- and DOM-sulfate. The gradient of the short routine method was: 0.0–0.5 min: 10% B, 6.0 min: 90% B, 6.1–7.5 min: 100% B, 7.6–10.0 min: 10% B. Time segments of the gradient of the long method were: 0.0–0.5 min: 10% B, 4.5 min: 40% B, 8.0 min: 45% B, 8.5–10.9 min: 100% B, 11.0–13.5 min: 10% B. The injection volume was 3 µL and the LC eluent was diverted to the MS between 2.3 and 6.0 min (8.0 min for the long method).

Tandem mass spectrometric detection was performed in negative selected reaction monitoring (SRM) mode after electrospray ionization. The following ion source settings were used: Temperature 400 °C, ion spray voltage -4500 V, curtain gas 35 psi, ion source gas 1 80 psi, ion source gas 2 90 psi, collision gas (N_2) high. SRM parameters (declustering potential, collision energy, collision cell exit potential) were optimized for the individual analytes by software-controlled compound optimization and are listed in Supplementary Table S3. Analyst® software version 1.6.2 (Sciex) was used for instrument control and data evaluation.

4.6. Method Validation

Validation of the short routine method included determination of the apparent recovery (R_A), recovery of extraction (R_E), and mass spectrometric matrix effects (SSE) for excreta of turkeys and chickens as well as assessment of SSE for extracts of excreta and GI content samples of pullets, and excreta extracts of roosters. In addition, intra- and inter-day repeatability of sample work-up and analysis, limits of detection and quantitation, and linear range of calibration functions were evaluated. The long gradient method was validated with respect to SSE in excreta extracts of turkeys and chickens.

For assessment of R_A, R_E, and SSE in excreta samples of turkeys and chickens, 300 mg aliquots of freeze dried excreta collected before administration of toxins were spiked with 60 μL aliquots of spiking solutions containing between 0.5 and 300 mg/L of DON-3-sulfate, DOM-3-sulfate, DON, DOM, DONS 1, DONS 2, DONS 3, and DOMS 2, resulting in 0.1 to 60 mg/kg of these analytes in the freeze dried sample aliquots. Spiking was performed at seven concentration levels (0.1/0.3/1/3/10/30/60 mg/kg) in triplicate. One hour after spiking, 21 spiked and three unspiked samples were worked-up and diluted as described above, and measured in the same run as matrix-matched and pure solvent calibration functions. Pooled diluted extracts of unspiked samples were used to prepare matrix-matched calibration functions as described in detail in [24]. Both matrix-matched and pure solvent calibration functions were established at seven concentration levels (1.1/3.4/11/34/112/337/674 ng/mL), which corresponded to the theoretical analyte concentrations in measurement solutions of samples spiked prior to work-up in the case of 100% apparent recovery. R_A, R_E, and SSE were calculated by comparing the slopes of the standard addition curve (k_{SA}), matrix-matched calibration curve (k_{MM}), and pure solvent calibration curve (k_{SOL}) as described in [41]. The following equations were used: $R_A = k_{SA}/k_{SOL} \times 100$; $R_E = k_{SA}/k_{MM} \times 100$;

SSE = $k_{MM}/k_{SOL} \times 100$. Matrix-matched calibration functions were established between 3 and 600 ng/mL in diluted extracts of chyme and excreta samples of the negative control group of pullets and in excreta sample extracts of roosters, and used for calculation of SSE in these matrices. Apparent recoveries in the same matrices were estimated by multiplying the SSE of each individual analyte by the average R_E of the same analyte determined in excreta of turkeys and chickens (see above).

The intra-day repeatability was determined by triplicate work-up and measurement of blank turkey and chicken excreta samples spiked at four concentration levels (1, 3, 10, 30 ng/g) on one day. The inter-day repeatability was assessed by work-up and analysis of two randomly selected excreta samples (one of turkey, one of chicken) on each day sample preparation and analysis was carried out.

Limits of detection (LOD, signal to noise ratio (S/N) 3/1) and quantitation (LOQ, S/N 10/1) were determined in pure solvent standard solutions and in matrix-matched standard solutions. LODs and LOQs in freeze-dried samples were calculated by dividing LODs and LOQs in matrix-matched solutions by the recovery of extraction and by multiplying by the dilution factor. The linear range in matrix-matched calibration curves defined the upper end of the working range.

4.7. Design of the Feeding Trials

In total, excreta samples from four different feeding trials with poultry were analyzed. The parameters of the individual feeding trials are summarized in Table 5. Turkeys, chickens, and pullets of the negative control groups were fed with basal poultry feed naturally contaminated with 0.2–0.3 mg/kg DON. Turkeys and chickens of the DON group received basal poultry feed enriched with DON from culture material to a concentration of 1.5–1.7 mg/kg. Similarly, turkeys and chickens of the DOM group were fed with basal poultry feed artificially contaminated with the equimolar concentration of DOM. Pullets were exposed to 4.4 mg/kg DON in feed. As this feed was prepared from wheat contaminated with different Fusarium toxins [28], it also contained traces of 3-acetyl-DON (0.13 mg/kg) and 15-acetyl-DON (0.03 mg/kg). Roosters received diet containing 11 mg/kg DON.

All feeding trials and animal experiments were conducted following the European Guidelines for the Care and Use of Animals for Research Purpose [42]. The feeding trials with turkeys and chickens lasted for one day during which animals had unlimited access to feed and water until feed was removed in the evening. Animals were housed under a light/dark cycle of 18/6 h. Feed consumption of the individual animals was recorded. Excreta samples were collected at regular intervals during the day and in the morning of the following day. After the experiment, animals received basal poultry feed for two weeks before they were reintegrated into the meat production process. In the animal experiment with pullets, pullets received 65 g feed/day in two equal portions for two weeks and excreta samples were collected twice a day in the second week [28]. Pooled excreta samples were used for determination of biological recoveries. On the last day of the trial, pullets were slaughtered 3 h after the last feeding and chyme samples (contents of jejunum and ileum) were taken. Roosters were fed the DON contaminated feed at amounts of 90 g/day and excreta were collected for seven days [30]. Samples taken on different days were pooled for each animal prior to analysis. All excreta and chyme samples were stored frozen until lyophilization. Freeze-dried samples were stored at −20 °C.

Table 5. Overview of four DON/DOM feeding trials with poultry.

	Turkeys	Chickens	Pullets	Roosters
Reference	-	-	[28]	[29,30]
Animals	Hybrid Converter	Ross 308	Lohmann LSL	New Hampshire hybrids
Age	11 weeks	5 weeks	12 weeks	adult
No. of animals/group	4	4	9	8
c (DON) in feed (DON group) (mg/kg)	1.5	1.7	4.4	11
c (DON, DOM) in feed (DOM group) (mg/kg)	DON: 0.3 DOM: 1.4	DON: 0.2 DOM: 1.6	-	-
c (DON) in feed negative control group (mg/kg)	0.3	0.2	0.2	No NC
Feeding	Ad libitum access from 7:00 to 22:00	Ad libitum access from 7:00 to 19:00	65 g feed/day (2 equal portions)	90 g feed /day
Duration	1 day	1 day	2 weeks	9 days *
Sampling times	Every 2 h until 15:00; 18:00; 22:00; 7:00 next day	Every 3 h until 19 h; 7:00 next day	Excreta: morning and afternoon for 1 week; chyme: 3 h after administration	Morning and afternoon for 1 week
Samples taken	Excreta	Excreta	Excreta, content of jejunum and ileum	Excreta

* The duration of the animal experiment was 10 months, but the period of restricted feeding lasted for nine days.

4.8. Analysis of Samples and Data Evaluation

Excreta and GI content samples were worked-up in duplicate, each pooled and diluted extract was measured once. Analyte concentrations in freeze-dried samples were determined on the basis of pure solvent calibration functions (peak area *versus* analyte concentration) established routinely between 1 and 300 ng/mL under consideration of the apparent recoveries. In case the concentrations determined by duplicate sample work-up and analysis differed by more than 20%, work-up and analysis was repeated in duplicate and the new values were taken for further data processing.

For determination of the biological recoveries of ingested DON and DOM in the trials with turkeys, chickens, and pullets, DON equivalent amounts of all DON- and DOM metabolites quantified in excreta were calculated and divided by the DON equivalent amounts of DON or DOM ingested in the corresponding time period (one day for turkeys and chickens, one week for pullets). Statistical evaluation of differences in metabolite concentrations in excreta samples of pullets with and without *A. galli* infection was performed by one-way analysis of variance in MS Excel 2013.

4.9. Toxicity Assessment of DOM-3-Sulfate and DOM-15-Sulfate

To determine the *in vitro* toxicities of DON, DOM, DOM-3-sulfate, and DOM-15-sulfate, an *in vitro* transcription/translation assay with rabbit reticulocyte lysate (Promega, Madison, WI, USA) was employed. Transcription/translation reactions were carried out according to the manufacturer's instructions as described in [33] for wheat germ extract with two modifications. Firstly, animal ribosomes (rabbit reticulocyte lysate) were used instead of wheat ribosomes. Secondly, the reactions were stopped after 20 min instead of 30 min. All substances tested were dissolved in water prior to use. The concentrations of DON in the assay were between 0 and 10 μM, while the concentration range of DOM and DOM-sulfates was 0–500 μM. For each compound, three independent assays were performed on three different days using the same batch of the reticulocyte lysate. For determining the translation efficiency at different inhibitor concentrations, one single test reaction was performed for each inhibitor at each concentration level. In addition, a control reaction was carried out at each concentration level where water was used instead of the test substance. The luciferase activity determined for the control reaction was set to 100% (uninhibited reaction) and the readouts from the other reactions were related to this control. Statistical evaluation was performed with IBM SPSS Statistics software. Comparison of mean values was performed with a two-sided *T*-test with Welch correction (equal variances not assumed). Results were considered significant at $p < 0.05$.

5. CONCLUSIONS

An LC-MS/MS based method for quantitative determination of DON-3-sulfate, DOM-3-sulfate, DON, DOM, DON sulfonates 1, 2, 3, and DOM sulfonate 2 in excreta and chyme samples of poultry was developed and validated. Application of the method to excreta and chyme samples from four different feeding trials with turkeys, broiler chickens, pullets, and roosters confirmed DON-3-sulfate as major DON metabolite in all investigated poultry species. Orally administered DOM was equally extensively metabolized to its 3-sulfate metabolite. In addition, DOM-3-sulfate was shown to be an important natural DON metabolite in pullets and roosters, where it amounted to *ca.* 12% and 33%, respectively, of all the detected metabolites. Interestingly, pullets not infected with *A. galli* excreted significantly greater concentrations and greater amounts of DON-3-sulfate and DOM-3-sulfate than pullets infected with worms. DON sulfonates of the series 1, 2 and 3 were detected in chyme samples of broilers, albeit only at trace levels. Biological recoveries of orally administered DON in the form of DON-3-sulfate and DOM-3-sulfate in chickens were close to 100%, supporting the hypothesis of nearly quantitative conversion of DON to its 3-sulfate metabolites in chickens. Although turkeys showed very similar metabolization of DON and DOM as chickens, biological recoveries of orally administered DON and DOM were only between 50% and 70%, suggesting faster excretion of DON and DOM metabolites in chickens than in turkeys.

Similar to DON-3-sulfate, DOM-3-sulfate was much less toxic than DON on the ribosome. Although *in vitro* transcription/translation assays cannot predict all aspects of *in vivo* toxicity, greatly reduced toxicity on the ribosome, the molecular target of trichothecenes, and low susceptibility of poultry to DON indicates that sulfation serves as detoxification mechanism for DON in poultry. Concluding from literature reports and our current study, locations of formation of DON-3-sulfate are most likely the intestinal mucosa, liver, and possibly the kidney. Elimination into excreta is probably achieved via bile and via urine. DOM-3-sulfate might also be formed by microbial conversion of DON-3-sulfate after retrograde movement of urinary DON-3-sulfate into the microbial rich caecae. In order to confirm this hypothesis, more complex animal experiments with separate collection of blood (hepatic and renal portal vein blood, systemic blood), bile, urine, contents of jejunum, ileum, caecae, and cloaca samples are required which demands highly sophisticated methods of sampling.

ACKNOWLEDGMENTS

The authors would like to thank the Federal Ministry of Science, Research and Economy, the National Foundation for Research, Technology and Development and BIOMIN Holding GmbH for funding the Christian Doppler Laboratory for Mycotoxin Metabolism. Work of Philipp Fruhmann and Gerlinde Wiesenberger was funded by the Vienna Science and Technology Fund (WWTF LS12-021) and the Austrian Science Fund (SFB Fusarium F3702 and F3706). In addition, we are grateful to Marco Reiter and Emanuel Ott for carrying out sample preparations.

Furthermore, we would like to thank Dieter Moll and Verena Klingenbrunner for providing a crude preparation of DOM and Veronika Slavik for preparative isolation of DOM. Finally, we are grateful to Nicole Reisinger for help with statistical data analysis.

AUTHOR CONTRIBUTIONS

P.F. and J.W. synthesized the DOM-sulfates. P.F. carried out NMR analysis as well as NMR data evaluation and wrote one part of the manuscript. S.D. conceived and designed the feeding trials with pullets and roosters and provided valuable information for the discussion part of the manuscript. G.W. performed and evaluated the toxicity tests. H.E.S.-Z. and S.C. performed the lab analyses. H.E.S.-Z. analyzed the data, contributed to study design and wrote the manuscript. F.B. contributed to study design and data analysis. F.B., P.F., S.D., G.W. and J.W. carefully revised the manuscript.

CONFLICTS OF INTEREST

The authors declare no conflict of interest.

REFERENCES

1. Morooka, N.; Uratsuji, N.; Yoshizawa, T.; Yamamoto, H. Toxic substances in barley infected with fusarium species. *Shokuhin Eiseigaku Zasshi* **1972**, *13*, 368–375.

2. Streit, E.; Schatzmayr, G.; Tassis, P.; Tzika, E.; Marin, D.; Taranu, I.; Tabuc, C.; Nicolau, A.; Aprodu, I.; Puel, O.; Oswald, I.P. Current situation of mycotoxin contamination and co-occurrence in animal feed-focus on europe. *Toxins* **2012**, *4*, 788–809.

3. Turner, P.C. Deoxynivalenol and nivalenol occurrence and exposure assessment. *World Mycotoxin J.* **2010**, *3*, 315–321.

4. Pestka, J.J. Toxicological mechanisms and potential health effects of deoxynivalenol and nivalenol. *World Mycotoxin J.* **2010**, *3*, 323–347.

5. Karlovsky, P. Biological detoxification of the mycotoxin deoxynivalenol and its use in genetically engineered crops and feed additives. *Appl. Microbiol. Biotechnol.* **2011**, *91*, 491–504.

6. Awad, W.A.; Ghareeb, K.; Bohm, J.; Zentek, J. Decontamination and detoxification strategies for the *Fusarium* mycotoxin deoxynivalenol in animal feed and the effectiveness of microbial biodegradation. *Food Addit. Contam: Part. A* **2010**, *27*, 510–520.

7. Berthiller, F.; Crews, C.; Dall'Asta, C.; Saeger, S.D.; Haesaert, G.; Karlovsky, P.; Oswald, I.P.; Seefelder, W.; Speijers, G.; Stroka, J. Masked mycotoxins: A review. *Mol. Nutr. Food Res.***2013**, *57*, 165–186.

8. Warth, B.; Sulyok, M.; Berthiller, F.; Schuhmacher, R.; Krska, R. New insights into the human metabolism of the *Fusarium* mycotoxins deoxynivalenol and zearalenone. *Toxicol. Lett.* **2013**,*220*, 88–94.

9. Dänicke, S.; Brezina, U. Kinetics and metabolism of the *Fusarium* toxin deoxynivalenol in farm animals: Consequences

for diagnosis of exposure and intoxication and carry over. *Food Chem. Toxicol.* **2013**, *60*, 58–75.

10. Wu, Q.; Dohnal, V.; Huang, L.; Kuca, K.; Yuan, Z. Metabolic pathways of trichothecenes.*Drug Metab. Rev.* **2010**, *42*, 250–267.

11. Maul, R.; Warth, B.; Schebb, N.H.; Krska, R.; Koch, M.; Sulyok, M. *In vitro* glucuronidation kinetics of deoxynivalenol by human and animal microsomes and recomninant human UGT enzymes. *Arch. Toxicol.* **2015**, *89*, 949–960.

12. Maul, R.; Warth, B.; Kant, J.S.; Schebb, N.H.; Krska, R.; Koch, M.; Sulyok, M. Investigation of the hepatic glucuronidation pattern of the Fusarium mycotoxin deoxynivalenol in various species. *Chem. Res. Toxicol.* **2012**, *25*, 2715–2717.

13. Uhlig, S.; Ivanova, L.; Faeste, C.K. Enzyme-assisted synthesis and structural characterization of the 3-, 8-, and 15-glucuronides of deoxynivalenol. *J. Agric. Food Chem.* **2013**, *61*, 2006–2012.

14. Nagl, V.; Wöchtl, B.; Schwartz-Zimmermann, H.E.; Hennig-Pauka, I.; Moll, W.D.; Adam, G.; Berthiller, F. Metabolism of the masked mycotoxin deoxynivalenol-3-glucoside in pigs.*Toxicol. Lett.* **2014**, *229*, 190–197.

15. Goyarts, T.; Danicke, S. Bioavailability of the *Fusarium* toxin deoxynivalenol (DON) from naturally contaminated wheat for the pig. *Toxicol. Lett.* **2006**, *163*, 171–182.

16. Sundstol Eriksen, G.; Pettersson, H.; Lindberg, J.E. Absorption, metabolism and excretion of 3-acetyl DON in pigs. *Arch. Anim. Nutr.* **2003**, *57*, 335–345.

17. Turner, P.C.; Hopton, R.P.; Lecluse, Y.; White, K.L.; Fisher, J.; Lebailly, P. Determinants of urinary deoxynivalenol and de-epoxy deoxynivalenol in male farmers from Normandy, France. *J. Agric. Food Chem.* **2010**, *58*, 5206–5212.

18. Turner, P.C.; Hopton, R.P.; White, K.L.; Fisher, J.; Cade, J.E.; Wild, C.P. Assessment of deoxynivalenol metabolite profiles in UK adults. *Food Chem. Toxicol.* **2011**, *49*, 132–135

19. Warth, B.; Sulyok, M.; Fruhmann, P.; Berthiller, F.; Schuhmacher, R.; Hametner, C.; Adam, G.; Frohlich, J.; Krska, R. Assessment of human deoxynivalenol exposure using an LC-MS/MS based biomarker method. *Toxicol. Lett.* **2012**, *211*, 85–90.

20. Danicke, S.; Brussow, K.P.; Valenta, H.; Ueberschar, K.H.; Tiemann, U.; Schollenberger, M. On the effects of graded levels of *Fusarium* toxin contaminated wheat in diets for gilts on feed intake, growth performance and metabolism of deoxynivalenol and zearalenone. *Mol. Nutr. Food Res.* **2005**, *49*, 932–943.

21. Winkler, J.; Kersten, S.; Valenta, H.; Huether, L.; Meyer, U.; Engelhardt, U.; Dänicke, S. Simultaneous determination of zearalenone, deoxynivalenol and their metabolites in bovine urine as biomarkers of exposure. *World Mycotoxin J.* **2015**, *8*, 63–74.

22. Seeling, K.; Danicke, S.; Valenta, H.; Van Egmond, H.P.; Schothorst, R.C.; Jekel, A.A.; Lebzien, P.; Schollenberger, M.; Razzazi-Fazeli, E.; Flachowsky, G. Effects of *Fusarium* toxin-contaminated wheat and feed intake level on the biotransformation and carry-over of deoxynivalenol in dairy cows. *Food Addit. Contam.* **2006**, *23*, 1008–1020.

23. Nagl, V.; Schwartz, H.; Krska, R.; Moll, W.D.; Knasmuller, S.; Ritzmann, M.; Adam, G.; Berthiller, F. Metabolism of the masked mycotoxin deoxynivalenol-3-glucoside in rats.*Toxicol. Lett.* **2012**, *213*, 367–373.

24. Schwartz-Zimmermann, H.E.; Hametner, C.; Nagl, V.; Slavik, V.; Moll, W.D.; Berthiller, F. Deoxynivalenol (DON) sulfonates as major DON metabolites in rats: From identification to biomarker method development, validation and application. *Anal. Bioanal. Chem.* **2014**, *406*, 7911–7924.

25. Wan, D.; Huang, L.; Pan, Y.; Wu, Q.; Chen, D.; Tao, Y.; Wang, X.; Liu, Z.; Li, J.; Wang, L.; Yuan, Z. Metabolism, distribution, and excretion of deoxynivalenol with combined techniques of radiotracing, high-performance liquid chromatography ion trap time-of-flight mass spectrometry, and online radiometric detection. *J. Agric. Food Chem.* **2014**, *62*, 288–296.

26. Devreese, M.; Antonissen, G.; Broekaert, N.; De Mil, T.; De Baere, S.; Vanhaecke, L.; De Backer, P.; Croubels, S. Toxicokinetic study and oral bioavailability of DON in turkey poults, and comparative biotransformation between broilers and turkeys. *World Mycotoxin J.***2015**, *8*, 533–539.

27. Yunus, A.W.; Valenta, H.; Abdel-Raheem, S.M.; Döll, S.; Dänicke, S.; Böhm, J. Blood plasma levels of deoxynivalenol and its de-epoxy metabolite in broilers after a single oral dose of the toxin. *Mycotox Res.* **2010**, *26*, 217–220.

28. Dänicke, S.; Beinekeb, A.; Rautenschlein, S.; Valenta, H.; Kersten, S.; Gaulyd, M. *Ascaridia galli* infection affects pullets differently when feed is contaminated with the *Fusarium* toxin deoxynivalenol (DON). *Vet. Parasitol.* **2013**, *198*, 351–363.

29. Ebrahem, M.; Kersten, S.; Valenta, H.; Breves, G.; Beineke, A.; Hermeyer, K.; Dänicke, S. Effects of feeding deoxynivalenol (DON)-contaminated wheat to laying hens and roosters of different genetic background on the reproductive performance and health of the newly hatched chicks. *Mycotox Res.* **2014**, *30*, 131–140.

30. Ebrahem, M.; Kersten, S.; Ehling, C.; Valenta, H.; Taylor, U.; Baulain, U.; Breves, G.; Beineke, A.; Hermeyer, K.; Dänicke, S. Effect of Fusarium toxin contaminated wheat on health, nutrient digestibility and semen quality of adult cockerels. *Europ. Poult. Sci.* **2014**, *78*, 1–17.

31. Fruhmann, P.; Skrinjar, P.; Weber, J.; Mikula, H.; Warth, B.; Sulyok, M.; Krska, R.; Adam, G.; Rosenberg, E.; Hametner, C.; Fröhlich, J. Sulfation of deoxynivalenol, its acetylated derivatives, and T2-toxin. *Tetrahedron* **2014**, *70*, 5260–5266.

32. Schwartz-Zimmermann, H.E.; Paulick, M.; Dänicke, S.; Schatzmayr, D.; Berthiller, F. Determination of deoxynivalenol sulfonates in cereal samples: Method development, validation and application. *World Mycotoxin J.* **2014**, *7*, 233–245.

33. Warth, B.; Fruhmann, P.; Wiesenberger, G.; Kluger, B.; Sarkanj, B.; Lemmens, M.; Hametner, C.; Frohlich, J.; Adam, G.; Krska, R.; Schuhmacher, R. Deoxynivalenol-sulfates: Identification and quantification of novel conjugated (masked) mycotoxins in wheat. *Anal. Bioanal. Chem.***2015**, *407*, 1033–1039.

34. Schwartz, H.E.; Hametner, C.; Slavik, V.; Greitbauer, O.; Bichl, G.; Kunz-Vekiru, E.; Schatzmayr, D.; Berthiller, F. Characterization of three deoxynivalenol sulfonates formed by reaction of deoxynivalenol with sulfur reagents. *J. Agric. Food Chem.* **2013**, *61*, 8941–8948.

35. Awad, W.A.; Hess, M.; Twarużek, M.; Grajewski, J.; Kosicki, R.; Böhm, J.; Zentek, J. The impact of the *Fusarium* mycotoxin deoxynivalenol on the health and performance of broiler chickens. *Int. J. Mol. Sci.* **2011**, *12*, 7996–8012.

36. Rotter, B.; Prelusky, D.B. Toxicology of deoxynivalenol (vomitoxin). *J. Toxicol. Environ. Health* **1996**, *48*, 1–34.

37. Prelusky, D.B.; Hamilton, R.M.G.; Trenholm, H.L.; Miller, J.D. Tissue distribution and excretion of radioactivity following administration of 14C-labeled deoxynivalenol to white leghorn hens. *Fundam. Appl. Toxicol.* **1986**, *7*, 635–645.

38. Lun, A.K.; Moran, E.T., Jr.; Young, L.G.; McMillan, E.G. Disappearance of deoxynivalenol from digesta progressing along the chicken's gastrointestinol tract after intubation with feed containing contaminated corn. *Bull. Environ. Contam. Toxicol.* **1988**, *40*, 317–324.

39. Lun, A.K.; Moran, E.T., Jr.; Young, L.G.; McMillan, E.G. Absorption and elimination of an oral dose of 3H-deoxynivalenol in colostomized and intact chickens. *Bull. Environ. Contam. Toxicol.* **1989**, *42*, 919–925.

40. Osselaere, A.; Devreese, M.; Goossens, J.; Vandenbroucke, V.; De Baere, S.; De Backer, P.; Croubels, S. Toxicokinetic study and absolute oral bioavailability of deoxynivalenol, T-2 toxin and zearalenone in broiler chickens. *Food Chem. Toxicol.* **2013**, *51*, 350–355.

41. Sulyok, M.; Krska, R.; Schuhmacher, R. A liquid chromatography/tandem mass spectrometric multi-mycotoxin method for the quantification of 87 analytes and its application to semi-quantitative screening of moldy food samples. *Anal. Bioanal. Chem.* **2007**,*389*, 1505–1523.

42. Official Journal of the European Union, Directive 2010/63/EU of the European Parliament and of the Council of 22 September 2010 on the Protection of Animals Used for Scientific Purposes. Available online: http://faolex.fao.org/docs/ pdf/eur 98296.pdf (accessed on 4 November 2015).

Index